Post-Transcriptional Regulation by STAR Proteins

ADVANCES IN EXPERIMENTAL MEDICINE AND BIOLOGY

Recent Volumes in this Series

Volume 685
DISEASES OF DNA REPAIR
Edited by Shamim I. Ahmad

Volume 686
RARE DISEASES EPIDEMIOLOGY
Edited by Manuel Posada de la Paz and Stephen C. Groft

Volume 687
BCL-2 PROTEIN FAMILY: ESSENTIAL REGULATORS OF CELL DEATH
Edited by Claudio Hetz

Volume 688
SPHINGOLIPIDS AS SIGNALING AND REGULATORY MOLECULES
Edited by Charles Chalfant and Maurizio Del Poeta

Volume 689
HOX GENES: STUDIES FROM THE 20TH TO THE 21ST CENTURY
Edited by Jean S. Deutsch

Volume 690
THE RENIN-ANGIOTENSIN SYSTEM: CURRENT RESEARCH PROGRESS
IN THE PANCREAS
Edited by Po Sing Leung

Volume 691
ADVANCES IN TNF FAMILY RESEARCH
Edited by David Wallach

Volume 692
NEUROPEPTIDE SYSTEMS AS TARGETS FOR PARASITE AND PEST CONTROL
Edited by Timothy G. Geary and Aaron G. Maule

Volume 693
POST-TRANSCRIPTIONAL REGULATION BY STAR PROTEINS
Edited by Talila Volk and Karen Artzt

Post-Transcriptional Regulation by STAR Proteins

Control of RNA Metabolism in Development and Disease

Edited by

Talila Volk, BSc, MSc, PhD
Department of Molecular Genetics, Weizmann Institute of Science, Rehovot, Israel

Karen Artzt, BA, PhD
Department of Molecular Genetics and Microbiology, Institute of Cell and Molecular Biology, University of Texas at Austin, Austin, Texas, USA

Springer Science+Business Media, LLC

Landes Bioscience

Springer Science+Business Media, LLC
Landes Bioscience

Printed in the USA.

Springer Science+Business Media, LLC, 233 Spring Street, New York, New York 10013, USA
http://www.springer.com

Please address all inquiries to the publishers:
Landes Bioscience, 1002 West Avenue, Austin, Texas 78701, USA
Phone: 512/ 637 6050; FAX: 512/ 637 6079
http://www.landesbioscience.com

The chapters in this book are available in the Madame Curie Bioscience Database.
http://www.landesbioscience.com/curie

Post-Transcriptional Regulation by STAR Proteins: Control of RNA Metabolism in Development and Disease,
edited by Talila Volk and Karen Artzt. Landes Bioscience / Springer Science+Business Media, LLC dual
imprint / Springer series: Advances in Experimental Medicine and Biology.

ISBN: 978-1-4419-7004-6

Title page image: Two littermates, the right one is *quaking*.

Library of Congress Cataloging-in-Publication Data

Post-transcriptional regulation by STAR proteins : control of RNA metabolism in development and disease /
edited by Talila Volk, Karen Artzt.
    p. ; cm. -- (Advances in experimental medicine and biology ; v. 693)
 Includes bibliographical references and index.
 ISBN 978-1-4419-7004-6
 1. RNA-protein interactions. 2. RNA--Metabolism--Regulation. 3. Cellular signal transduction. I. Volk,
Talila. II. Artzt, Karen Jane. III. Series: Advances in experimental medicine and biology ; v. 693. 0065-2598
[DNLM: 1. RNA-Binding Proteins--metabolism. 2. Embryonic Development--physiology. 3. RNA Process-
ing, Post-Transcriptional. 4. Signal Transduction. 5. Transcriptional Activation. W1 AD559 v.693 2010 /
QU 55.2 P857 2010]
 QP623.8.P75P67 2010
 572.8'8--dc22
                                    2010021544

DEDICATIONS

This book is dedicated to my family. —Talila Volk

Dedicated to my mentors: Dorothea Bennett, 1929-1990 and L.C. Dunn, 1893-1974. —Karen Artzt

PREFACE

This book aims to bring to the forefront a field that has been developing since the late 1990s called the STAR pathway for Signal Transduction and Activation of RNA. It is a signaling pathway that targets RNA directly; in contrast to the canonical signal—kinase cascade—transcription factor—DNA—RNA. It is proposed to allow quick responses to environment changes such as those necessary in many biological phenomenona such as the nervous system, and during development. The pathway is diagramed in Chapter 1, Figure 1. This chapter is a historical introduction and general review with some new data on theoretical miRNAs binding sites and STAR mRNAs. In Chapter 2, Feng and Banks address the accumulating evidence that the RNA-binding activity and the homeostasis of downstream mRNA targets of STAR proteins can be regulated by phosphorylation in response to various extracellular signals. Then Ryder and Massi review the available information on the structure of the RNA binding STAR domain and provides insights into how these proteins discriminate between different RNA targets. Next Claudio Sette offers an overview of the post-translational modifications of STAR proteins and their effects on biological functions, followed by two chapters dedicated to in depth review of STAR function in spermatogenesis and in mammalian embryonic development. Chapters 7 and 8 discuss what can be learned from STAR proteins in non-mammalian species; in Drosophila and Gld-1 and Asd-2 in *C. elegans*. Next Rymond discusses the actual mechanics of splicing with mammalian SF1. Lastly Richard reviews what is known about STAR proteins and human disease including osteoporosis, schizophrenia, cancer, infertility and ataxia. The general intention of the editors is that basic researchers and clinicians will be stimulated to join the "Enterprise" studying the role of STAR proteins in other relevant diseases including dysmyelination and remyelination in multiple sclerosis and disorders of the neural and immune synapse.

Talila Volk, BSc, MSc, PhD
Karen Artzt, BA, PhD

ABOUT THE EDITORS...

TALILA VOLK is an Associate Professor in the field of Developmental Biology and the incumbent of the Sir Ernest B. Chain Professional Chair. Her major research interests are tissue morphogenesis and organogenesis during embryonic development. She has been studying the function and activity of the STAR family member Held Out Wing (HOW) in the fruit fly Drosophila since 1999. She served as the chairwoman for the Society of Developmental Biology in Israel (ISDB). Dr. Volk has gained her BSc from Tel-Aviv University, and her MSc and PhD degrees from the Weizmann Institute of Science, Rehovot, Israel.

ABOUT THE EDITORS...

KAREN ARTZT is an Ashbel Smith Professor Emeritus at the University of Texas at Austin where she directed a research laboratory for 20 years, where she was a member of the Section of Molecular Genetics and Microbiology. Prior to that she was an associate Member of the Memorial Sloan Kettering Cancer Center in New York. Her main research interests include developmental genetics with an emphasis on cancer biology. In collaboration with Tom Ebersole she identified and cloned the mouse gene *quaking* that was one of the founding members of the STAR family. Dr. Artzt received her academic degrees from Cornell University; a BA from the Ithaca campus and a PhD from the Medical College School of Graduate Sciences in New York City. In 1972 she spent a year as a Postdoctoral Fellow at the Pasteur Institute in Paris under the direction of the Nobel Prize winner, Francois Jacob.

PARTICIPANTS

Karen Artzt
Department of Molecular Genetics
  and Microbiology
Institute of Cell and Molecular Biology
University of Texas at Austin
Austin, Texas
USA

Andrew Bankston
Feng Laboratory
Department of Pharmacology
Emory University School of Medicine
Atlanta, Georgia
USA

Ingrid Ehrmann
Institute of Human Genetics
Newcastle University
International Centre for Life
Newcastle
UK

David J. Elliott
Institute of Human Genetics
Newcastle University
International Centre for Life
Newcastle
UK

Yue Feng
Feng Laboratory
Department of Pharmacology
Emory University School of Medicine
Atlanta, Georgia
USA

Karen K. Hirschi
Department of Pediatrics
Baylor College of Medicine
Houston, Texas
USA

Monica J. Justice
Departments of Molecular and Human
  Genetics and Molecular Physiology
  and Biophysics
Baylor College of Medicine
Houston, Texas
USA

Min-Ho Lee
Department of Biological Sciences
University at Albany
State University of New York
Albany, New York
USA

Francesca Massi
Department of Biochemistry and Molecular
  Pharmacology
University of Massachusetts Medical
  School
Worcester, Massachusetts
USA

Stéphane Richard
Terry Fox Molecular Oncology Group
Bloomfield Center for Research on Aging
Lady Davis Institute for Medical Research
Departments of Oncology and Medicine
McGill University
Montréal, Québec
Canada

Sean P. Ryder
Department of Biochemistry and Molecular
  Pharmacology
University of Massachusetts Medical
  School
Worcester, Massachusetts
USA

Brian C. Rymond
Biology Department
University of Kentucky
Lexington, Kentucky
USA

Tim Schedl
Department of Genetics
Washington University School of Medicine
St. Louis, Missouri
USA

Claudio Sette
Department of Public Health and Cell
  Biology
University of Rome "Tor Vergata"
and
Institute for Neuroscience IRCSS
  Fondazione Santa Lucia
Rome
Italy

Talila Volk
Department of Molecular Genetics
Weizmann Institute of Science
Rehovot
Israel

Jiang I. Wu
Department of Physiology
  and Developmental Biology
University of Texas Southwestern
  Medical Center
Dallas, Texas
USA

CONTENTS

1. STAR TREK: AN INTRODUCTION TO STAR FAMILY PROTEINS AND REVIEW OF QUAKING (QKI)...1

Karen Artzt and Jiang I. Wu

Abstract..1
History of the STAR Family..1
The Domain Structure and Alternate Splicing of STAR Proteins.......................4
STAR Proteins Have a Multitude of Developmental Functions 5
Diverse Molecular Functions of STAR Proteins in RNA Processing..................5
Qk Expression in the Adult Nervous System and Disease.................................6
Qk 3′ UTR Conservation and a High Theoretical Number of miRNA Binding Sites............8
Discussion and Conclusion 11
Future Applications, New Research, Anticipated Developments 21

2. THE STAR FAMILY MEMBER: QKI AND CELL SIGNALING....................25

Yue Feng and Andrew Bankston

Abstract..25
Introduction..25
QKI Is Essential for Embryonic and Postnatal Development 26
Phosphorylation of QKI Isoforms by Src-PTKS Regulates the Cellular Fate
    of QKI mRNA Targets at Multiple Post-Transcriptional Steps......................27
Numerous Extracellular Signals Can Be Linked to the Src-PTK-QKI Pathway 30
Potential Role of QKI And Src-PTK Signaling in Tumorigenesis
    and Cognitive Diseases 32
Conclusion 33

3. INSIGHTS INTO THE STRUCTURAL BASIS OF RNA
RECOGNITION BY STAR DOMAIN PROTEINS...................................37

Sean P. Ryder and Francesca Massi

Abstract...37
Introduction..37
The STAR Domain 39
RNA Recognition by STAR Proteins...39
Star Domain Structure 43
Conclusion 50
Note Added in Proof...50

4. POST-TRANSLATIONAL REGULATION OF STAR PROTEINS
AND EFFECTS ON THEIR BIOLOGICAL FUNCTIONS 54

Claudio Sette

Abstract...54
Introduction..55
Sam68: A Brief Overview 55
Regulation of Sam68 Functions by Tyrosine Phosphorylation.........57
Regulation of Sam68 Functions by Serine/Threonine Phosphorylation.........59
Regulation of Sam68 Functions by Methylation...............................60
Regulation of Sam68 Functions by Acetylation and Sumoylation......61
Post-Translational Modifications of SLM-1 and SLM-2.....................61
Post-Translational Modifications of the QKI Proteins 62
Post-Translational Modifications of SF1..63
Conclusion 63

5. EXPRESSION AND FUNCTIONS OF THE STAR PROTEINS Sam68
AND T-STAR IN MAMMALIAN SPERMATOGENESIS.......................67

Ingrid Ehrmann and David J. Elliott

Abstract..67
Gene Expression Control in Spermatogenesis...................................67
Expression of STAR Proteins during Spermatogenesis......................70
Protein Structure and Modifications...70
Mouse Knockout Models Define the Roles of STAR Proteins in Testis Function 76
The STAR Protein Sam68 Is Involved in Translational Control in Spermatogenesis.........76
STAR Proteins Might Play Roles in Pre-mRNA Splicing Control in Spermatogenesis 77
Other Potential Roles of STAR Proteins in Spermatogenesis.............78
Conclusion 78

6. THE ROLE OF QUAKING IN MAMMALIAN EMBRYONIC
DEVELOPMENT 82

Monica J. Justice and Karen K. Hirschi

Abstract...82
Introduction...83

Quaking Is Required for the Formation of Embryonic Vasculature 84
QKI5 Regulates *QKI6* and *QKI7* in Visceral Endoderm 84
Molecular Basis of Blood Vessel Formation 85
Quaking Is Required for Visceral Endoderm Differentiated Function 86
Other Possible Roles for Quaking in Cardiovascular Development 88
The Evolving Roles of Quaking Function 88
Conclusion 89

7. *DROSOPHILA* STAR PROTEINS: WHAT CAN BE LEARNED FROM FLIES? 93

Talila Volk

Abstract 93
STAR Proteins in *Drosophila* 93
HOW Regulates Differentiation of Diverse Tissues 94
HOW and Kep1 Regulate Cell Division and Apoptosis in *Drosophila* 100
Conclusion 103
Note Added in Proof 104

8. *C. ELEGANS* STAR PROTEINS, GLD-1 AND ASD-2, REGULATE SPECIFIC RNA TARGETS TO CONTROL DEVELOPMENT 106

Min-Ho Lee and Tim Schedl

Abstract 106
Multiple Functions of GLD-1 in Germline Development 106
GLD-1 Molecular Analysis 109
mRNA Targets: GLD-1 Is a Translational Repressor 110
mRNA Targets: Further Insights into GLD-1 Function in Germline Development 114
mRNA Targets: Towards Defining the GLD-1 RNA Binding Motif and Mechanism
    of Translational Repression 115
How Is GLD-1 Expression Regulated? 117
ASD-2, Another *C. elegans* STAR Protein, Functions in Alternative Splicing 119
Conclusion 119

9. THE BRANCHPOINT BINDING PROTEIN: IN AND OUT OF THE SPLICEOSOME CYCLE 123

Brian C. Rymond

Abstract 123
BBP and SF1 Are Site-Specific RNA Binding Proteins 124
A BBP-Mud2 Heterodimer Functions in Branchpoint Recognition 126
BBP-MUD2 and the Dynamics of Early Spliceosome Assembly 127
Co-Transcriptional Pre-mRNA Splicing 130
But Is BBP Really an Essential Splicing Factor? 131
BBP Is Needed for the Nuclear Retention of Unprocessed Pre-mRNA 131
Uncoupling Pre-mRNA Splicing from the Synthesis of Functional mRNA 133
Does BBP Have a Cytoplasmic Function? 133
Does BBP Regulate the Fate of Intronless RNA? 134
Conclusion 135

10. Reaching for the STARS: Linking RNA Binding Proteins to Diseases..........142

Stéphane Richard

Abstract..142
Sam68: Its Discovery and Nomenclature...142
The KH Domain...143
Sam68 RNA Targets...145
Sam68 Cellular Localization..146
Sam68 Signaling Motifs..147
Arginine Methylation...148
STAR Protein Mouse Models...149
Sam68 Null Mice..149
QKI Mouse Models..150
STAR Proteins and Human Diseases..151
Osteoporosis..151
Schizophrenia..152
Ataxia...152
Cancer..152
Conclusion..153

INDEX...159

CHAPTER 1

STAR TREK
An Introduction to STAR Family Proteins and Review of Quaking (QKI)

Karen Artzt and Jiang I. Wu*

Abstract: The STAR family has an extremely diverse role during development and in RNA metabolism. We have concentrated on QKI as an example of this pleiotropic activity and also presented some new data on the role of its conserved 3'UTRs gleaned from bioinformatics analysis of theoretical miRNA binding sites. We review the concept of a direct pathway from signal transduction to activation of RNA, how this pathway could be the cell's quick response to developmental and physiological changes and how it must be tightly regulated.

HISTORY OF THE STAR FAMILY

The first member of the signal transduction and activation of RNA (STAR) family analyzed in detail was mammalian Src associated in mitosis (*SAM68* now also known as *KHDRBS1*).[1,2] It was identified for its role in transducing cell signals. *SAM68* was shortly joined by a subfamily of three conserved genes distinguished by their diverse and interesting mutant developmental phenotypes: the tumor suppressor gene *gld-1* in *C. elegans*,[3] the dysmyelinating gene *quaking* (*Qk*) in mouse[4] and a *Drosophila* gene *held out wings* (*how*) important for muscle development.[5,6] In 1996 the family was completed with a more distant relative human *splicing factor 1* (*SF1*)[7] (Fig. 1).

Later, additional members of the three subfamilies were characterized; among them were mammalian orthologs of *Sam68*: *Slm1* (now known as *Khdrbs2*) and *Slm2/T-Star* (now called *Khdrbs3*).[8,9] Very recently identified was *asd-2*,[10] (Table 1), a closer relative

*Corresponding Author: Jiang I. Wu—Department of Physiology and Developmental Biology, University of Texas Southwestern Medical Center, Dallas, TX 75390, USA.
Email: jiang9.wu@utsouthwestern.edu

Post-Transcriptional Regulation by STAR Proteins: Control of RNA Metabolism in Development and Disease, edited by Talila Volk and Karen Artzt.
©2010 Landes Bioscience and Springer Science+Business Media.

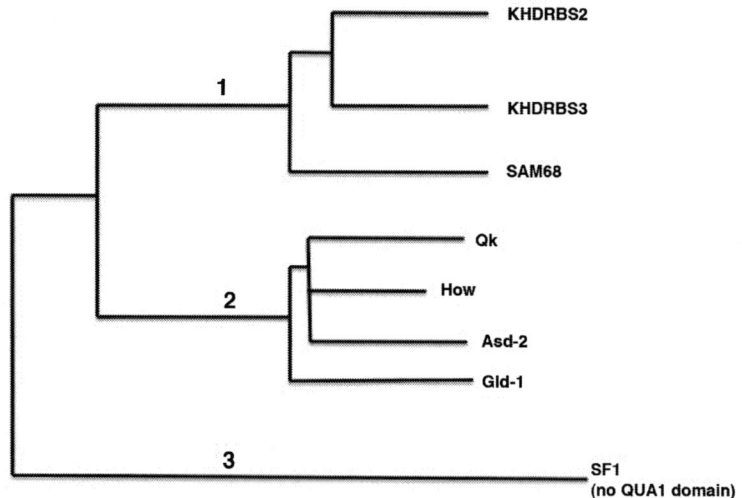

Figure 1. Highly simplified family of STAR proteins. The tree was redrawn from treefam accession numbers TF314878 and TF319159.[80] The three sub branches are numbered.

to *Qk* than *gld-1* in *C. elegans* based on its having three alternative spliced isoforms, two different 3′UTRs, a tyrosine tail and a closer phylogenic distance to *Qk* than *gld-1*. Members of STAR family are now defined from yeast to mammals and plants. The more distant relatives in plants have not been studied except for *SPL11-INTERACTING PROTEIN 1* (*SPIN1*) in rice. SPIN 1 negatively regulates programmed cell death and disease resistance. It is a member of the SF1 Family branch. In total, seven STAR paralogs including *SPIN1* were found in rice.[11] In some species it is hard to determine exact homologs because the species has expanded different family members; thus the blanks in Table 1. What is shared by all of the STAR proteins except SF1, is an uncommon single expanded KH RNA binding domain (Maxi-KH) that was flanked by two new conserved domains: an amino terminal QUA1 and a carboxyl terminal QUA2 domain. This triple domain

Table 1. STAR family members in different species

Mammals	Drosophila	C. elegans	Yeast	Plants
QKI	How	Asd-2 and Gld-1		AT1G09660, (Arabidopsis thaliana)
Sam68/Khdrbs1	Sam68 and Kep1			AT2G38610, (Arabidopsis thaliana)
T-star/Slm2/ Khdrbs3	qkr58E-1 or qkr54B (flymine)			
Slm1/Khdrbs2	qkr54B (flymine)			
SF1	SF1	sfa-1	BBP (pombe), MSL5 (cerevisiae)	SPIN1, rice (Oryza sativa), RIK [At5g51300] (Arabidopsis thaliana)

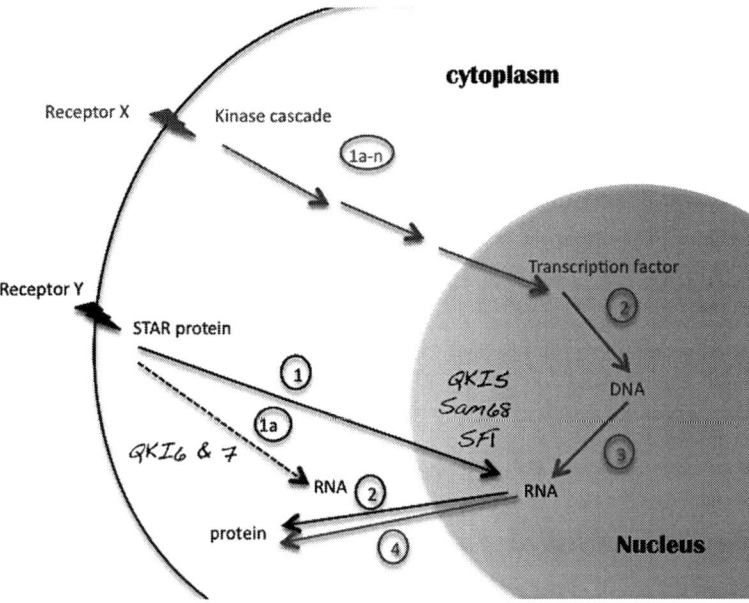

Figure 2. Theoretical schema of a more direct STAR pathway. Instead of the canonical pathways for signal transduction (red arrows) that requires at least four steps from receiving signals to protein production, the STAR pathway requires a minimum of two steps (blue arrows). QKI5, Sam68 and SF1 are for regulation of splicing and RNA metabolism in the nucleus; QKI6 and 7 are for RNA transport, stability and translational regulation in the cytoplasm. It is possible that heterodimers between different isoforms participate in some of the above functions. A color version of this image is available at www.landesbioscience.com/curie.

structure came to be called the STAR domain.[12] It is also known as the GSG domain.[3] The problem in the late 1990s was to make biological and molecular sense of this diverse and highly conserved family.

At the time, the most characterized family member was mammalian SAM68. By virtue of its KH domain, it was thought to bind RNA. There was also ample experimental evidence suggesting that it plays an important role in signal transduction because of its proline-rich regions, SH3- and WW-binding sites, RGG boxes and a prominent string of tyrosines in the C-terminal tail.[13,14] SAM68 is a substrate of SRC and FYN tyrosine kinases during mitosis.[1,2,15] Its tyrosine-phosphorylation can also be regulated by many signals and kinases including the activated insulin receptor and leptin receptor (refs. 13,16 and references within); additionally it is an ERK Ser/Thr kinase target.[17] Phosphorylation of SAM68 not only enables its interactions with many SH2/SH3- containing proteins to activate downstream signaling pathways, but also modulates its RNA regulating activities.[13,14] In addition to phosphorylation, post-translational modifications that regulate SAM68 function in RNA metabolism also include arginine methylation[18] and sumoylation.[19] All of the above suggest a direct connection between signal transduction and RNA regulation (Fig. 2). (See Feng and Bankston chapter for more detail.)

Because of the extreme conservation of QKI from *Drosophila* to mammals, it was noticeable that the tyrosine-rich tail was also present in the QKI subfamily members. Whereas little is conserved between the end of the STAR domain and the tyrosine tail, depending on which alignment is used, from 3 to 5 out of the 6 C-terminal tyrosines and their surrounding amino acids are identical between QKI and HOW.[12] In addition, QKI contains seven SH3-binding sites.[4] Thus in the late 1990s, based on all the phenotypic data provided by the STAR proteins, the functional analysis of SAM68 and the conserved protein structures, a link was hypothesized between signal transduction and some aspects of RNA regulation. Although STAR proteins differ in some of their features and participate in diverse steps of RNA metabolism, they were hypothesized to participate in a novel cellular process that brings external signals directly to RNA[12] (Fig. 2). Similar ideas about RNA-binding proteins as regulators of gene expression were also formulated by Siomi and Dreyfuss.[20]

Indeed, it did not take long to validate this notion for QKI. In 2003, Yue Feng's group found that Src family protein tyrosine kinases phosphorylate QKI at the C-terminal tyrosine cluster and modulates the ability of QKI to bind myelin basic protein (MBP) mRNA.[21] The same group later showed that it was FYN that phosphorylates QKI in brain (see Feng and Bankston chapter for details).[22] Similar to Sam68, QKI can also be arginine methylated.[18]

THE DOMAIN STRUCTURE AND ALTERNATE SPLICING OF STAR PROTEINS

STAR proteins differ from most other KH-containing RNA binding proteins in two ways. First, while most other such proteins have multiple KH domains, STAR proteins have only one maxi-KH domain with two extended loops and an additional C-terminal helix.[23,24] Since multiple KH domains are thought to stabilize RNA binding, STAR proteins may accomplish this by functioning as homo- and/or heterodimers.[25,26] Second, the central KH domain is flanked on either side by two domains called QUA1 and QUA2. Towards the amino terminal, the QUA1 domain is necessary for dimerization,[25,26] while the carboxyl side QUA2 is thought to participate in RNA binding.[23] The sole exception to this domain organization is SF1, which lacks the dimerizing QUA1 domain. Thus four out of the five mammalian STAR proteins have a central core of QUA1-KH-QUA2. The solution structures of the KH-QUA2 regions either from the SF1 in complex with RNA[23] or from *Xenopus* QKI in absence of RNA have been resolved.[24]

Another conserved aspect of STAR family genes is their own alternate splicing. For example, mammalian *Qk* and its *Drosophila* homologue *how* each have at least three major splice forms.[4,27] For *Qk*, this produces three different proteins QKI5, QKI6 and QKI7 so named for the length of their mRNAs. Each splice form produces its own unique carboxyl terminal.[4] These become a convenient handle for producing antibodies that distinguish the isoforms.[28] A fourth splice form for *Qk* has been identified (QKI7b), which also comes with its own carboxyl terminus.[29] It is these unique carboxyl tails that determine their locations in the cell. The QKI5 specific tail contains a nuclear localization signal and thus it is found mostly in the nucleus. However, QKI5 can also shuttle between the nucleus and the cytoplasm.[26] QKI6 and QKI7 have unique peptide tails and both reside in the cytoplasm exclusively.[28] Alternative splicing of the *Qk* gene also produces two distinctive long 3'UTRs, one used by QKI5 and one shared to different extents by QKI6 and QKI7.[4]

The different QKI splice forms also perform different functions; QKI5 is the major isoform in the embryo and is likely responsible for the early embryonic lethality of several ethylnitrosourea (ENU) induced mutations and the *Qk* knockout allele.[30-32] QKI6 and 7 are necessary for myelination as *Qk^viable* mice have a glial specific absence of QKI6 and QKI7 proteins and therefore develop the dysmyelination phenotype.[28,33] In fact in the Qk^{viable} allele, there is a 1 Mb deletion located less than 1000 bp 5' to Qk and includes not only the presumed glial cell enhancer for Qk but also part of the *Parkin 2* gene and *the Parkin co-regulated gene (Pacrg)*.[4,34] It is clear that while the glial cell enhancer is deleted in Qk^{viable}, the coding region necessary for embryo survival is intact. For a more detailed analysis of Qk's genomic structure and alternate splicing see reference 29.

STAR PROTEINS HAVE A MULTITUDE OF DEVELOPMENTAL FUNCTIONS

STAR proteins are widely expressed and have been shown by mouse mutations to function in various different developmental processes. The evidence for the essential developmental functions of QKI protein came from the studies of several spontaneous, ENU-induced or the knockout alleles of Qk. This implies tight developmental and temporal control of Qk function. Although many early embryonic lethal Qk alleles are not informative about later functions, the above notion is supported by the diversity of phenotypes generated by the genetic compounds of the many ENU induced alleles and one knockout allele. These include: abnormal somites, heart defects, cranial defects and a disorganized anterior-posterior axis,[31] lack of vascular development[35] and mis-regulation of visceral endoderm function.[36] Also on this list are: smooth muscle cell differentiation, kinky and open neural tubes.[32] The viable alleles of Qk point out its important function in the nervous system as the mutations lead to lack of myelination in both the central and peripheral nervous systems,[37] early-onset seizures, severe ataxia, a dramatically reduced lifespan and Purkinje cell axonal swellings indicative of neurodegeneration.[38] In addition, proportional dwarfism and ataxia have been documented in the compound of the Qk knockout and the Qk^{viable} alleles (Kuniya Abe, personal communication).

Surprisingly, considering Sam68's many roles in signal transduction and RNA regulation, the knockout mice have a relatively mild phenotype. This is possibly due to the presence of two highly related subfamily members SLM1 and SLM2.

Although Sam68 knockout mice are viable, they display male infertility[39] (see Erhmann and Elliot chapter), motor coordination defects,[40] resistance to age-related bone loss and reduced susceptibility to mammary gland tumors.[41,42]

Recently, SF1 has been shown to be involved in β-catenin/wnt pathway.[43] Heterozygous knockout mice display a higher incidence of drug induced colon tumors, which is possibly a wnt-dependent tumorigenesis process.[44] The homozygous *SF1* knockout mice die before embryonic day (E8.5), indicating SF1's function in splicing or other essential cellular processes in early development.[44]

Thus, it seems that STAR proteins regulate a multitude of functions at different times in various tissues in the embryo and adult. This diversity could be a function of the large number of target RNAs[45] and other regulatory mechanisms discussed below.

DIVERSE MOLECULAR FUNCTIONS OF STAR PROTEINS IN RNA PROCESSING

In addition to the above multitude of phenotypes, the molecular functions of STAR proteins and their orthologs in RNA metabolism are also diverse. STAR proteins have been shown to function in pre-mRNA splicing, mRNA localization and transport, mRNA stability and translation efficiency (46 and references within). In addition, many of the activities are regulated by extra cellular signals and post-translational modifications (see Sette chapter for detail).

The QKI target RNAs that have been most studied are the myelin set of genes. QKI5 affects pre-mRNA splicing of myelin-associated glycoprotein (MAG) via an alternative splicing element (QASE) in the downstream intron. In this instance QKI5 may function together with other splicing regulators to repress the developmentally regulated inclusion of MAG exon 12.[47] Interestingly, the QKI homologs, *Drosophila* HOW[48] and *C. elegans* ASD-2[10] have also been shown to regulate alternative splicing during development.

Studies of myelin basic protein (MBP) in *Qk^viable* mice demonstrated that QKI affects multiple steps of MBP mRNA metabolism. QKI7 and possibly QKI6 play a role in cytoplasmic stabilization of MBP mRNA presumably via the QKI RNA-binding element (QRE) in the MBP 3'-UTR.[49,50] QKI proteins have also been shown to regulate nuclear retention of MBP mRNA and transport to the myelinating membranes.[50,51] Among the several defined targets of QKI is the cyclin-dependent kinase (CDK)-inhibitor 1 (p27Kip1). Laroque et al showed that QKI6 and QKI7 promote oligodendrocyte differentiation possibly by binding and protecting p27Kip1 mRNA.[52] Similarly ectopically expressed QKI proteins promote Schwann cell differentiation and myelination.[53] In addition, QKI proteins regulate the stability of MAP1B mRNA and promote oligodendrocyte maturation.[54]

The role of STAR proteins in translation regulation first came from studies of the *C. elegans* GLD-1 protein. GLD-1 controls gene expression by acting through a hexanucleotide sequence (NACUCA) called TGE in the target 3'UTR to repress translation of Tra-1 protein (see Schdel and Lee chapter for more detail).[55] Later it was found that mouse QKI6 acts with a similar mechanism through the *Gli1* mRNA to repress translation.[56,57] The identified RNA targets of GLD-1 translational regulation includes *p53*, *Notch* and the *C. elegans caudal* homolog, *pal-1*.[58-60] It will be interesting to identify more RNA targets whose translations are regulated by STAR proteins.

The diverse molecular functions of SAM68 and SF1 will be discussed in detail in the later chapters.

Qk EXPRESSION IN THE ADULT NERVOUS SYSTEM AND DISEASE

The functions of QKI in neurological disease have recently attracted much attention. Although *Qk* is highly expressed in glia, it is not expressed in most neurons.[28] In fact during development, as embryonic neuroblasts differentiate into glia versus neurons, *Qk* is turned off in the neurons and up regulated in glia.[61] However, in the adult nervous system, according to the Allen mouse brain atlas,[62] *Qk* is abundantly expressed in Purkinje neurons in the cerebellum (Fig. 3). Also, the more severe viable *Qk ENU* mutation, *Qk^e5*, shows both a lack of myelination and Purkinje cell axonal swellings indicative of neurodegeneration and severe ataxia.[38] These recent data support the idea that QKI is part of an interaction network for human inherited ataxias and disorders of Purkinje cell degeneration.[63] In addition to the ataxia phenotype in *Qk^viable* mice, another connection of QKI to ataxias is that several highly expressed microRNAs in Purkinje cells that co-regulate Ataxin1 levels[64] also have frequent theoretical binding sites in the *Qk7* 3'UTR. They are: hsa-miR-19a, 19b, 101, 130a and 130b and have respectively 6, 5, 1, 2 and 2 theoretical sites in the *Qk7* 3'UTR(see below).

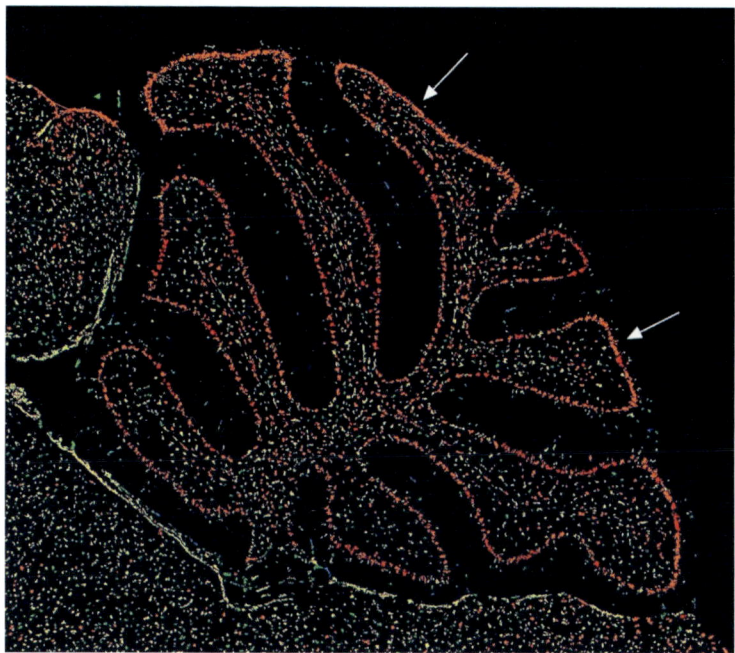

Figure 3. *Qk* expression in Purkinje neurons in the cerebellum. In situ hybridization results of *Qk* on a sagittal section of adult mouse brain (Allen Atlas Portal [Internet]. Seattle (WA): Allen Institute for Brain Science, ©2009). Red indicates the highest level of expression. White arrows point to the single row of Purkinje cells.

Another interesting but neglected place to look for a *Qk* disease connection is multiple sclerosis (MS). Relapsing-remitting MS is an autoimmune disease that results in cycles of demyelination and very occasional remyelination.[65] An increase in QKI during adult remyelination has been noted to happen in MS[66,67] but this idea has not been followed up. However, of the top four miRNAs most significantly dysregulated in MS[67a], three that are upregulated: hsa-miR-145, -186 and -664 and one that is downregulated hsa-miR-20b, there are respectively 1, 2, 5 and 6 predicted sites in the Qk 3'UTRs (see below).

Recently QKI has surfaced as a factor associated with schizophrenia, a disease accompanied by myelin deficiency. Decreased expression of the myelin-related genes, including QKI, is a consistent finding in studies of postmortem brain from subjects with schizophrenia.[68] Changes of QKI expression may be responsible for the impairment of myelin observed in schizophrenia[69] but as yet there are no polymorphic changes described in *Qk* in patients. Nevertheless, some genetic evidence has been reported implicating *Qk* as a susceptibility factor in schizophrenia.[70]

And lastly there is a very recent report of a human haploinsuffiency of the *QKI* gene with a phenotype reminiscent of 6q terminal deletion syndrome. The patient carried a balanced translocation with a breakpoint in *QKI* in the intron between exons 2 and 3. The symptoms included intellectual disabilities, hypotonia, seizures, brain anomalies, and specific dysmorphic features including short neck, broad nose with bulbous tip, large and low-set ears and downturned corners of the mouth.[70a]

Qk 3' UTR CONSERVATION AND A HIGH THEORETICAL NUMBER OF miRNA BINDING SITES

The diverse functions of STAR proteins in different cell types and developmental processes must require tight developmental and temporal control of their expression. For QKI, known for its regulation at the transcription and splicing levels, another yet potentially important level of regulation is by microRNAs (miRNAs) via its long and conserved 3'UTRs. The extreme conservation of the *QKI5* 3' UTR was initially noted by comparing mouse and *Xenopus Qk* (*Xqua*).[71] Surprisingly, one conserved block of over 300 nucleotides is 90% identical between *Xqua* and *QkI5*. In addition, most vertebrate *Qk* genes have two entirely different long 3'UTRs located some distance from each other in the genome. Both of them are highly conserved within vertebrate species (Fig. 4) and thus seemed likely targets for post-transcriptional regulation by miRNA. Using web-based bioinfomatic sources, we analyzed the potential miRNA binding sites in human *QKI* 3'UTRs. We chose humans since they have the highest number of miRNAs characterized.

To examine the predicted miRNA binding sites in the 3'UTRs of human *QKI*, we employed the microRNA.org database at Memorial Sloan-Kettering Cancer Center (http://www.microrna.org/microrna/home.do).[72] This database has been relatively recently updated (Sept 2008) to include a comprehensive cDNA sequencing project from a large set of mammalian tissues and cell lines. Thus it contains data for most if not all of 3'UTRs of a specific gene. This is not true for other databases that don't distinguish 3'UTRs and usually deal with only one. MicroRNA.org picked up *Qk* 3'UTRs of four different lengths: 6,399, 5,434, 4,173 and 3,512 bp respectively. By aligning these sequences it was evident that the first and fourth of these represent the unique and non-overlapping *Qk* 3'UTRs. With some genes, when added together the predicted miRNA binding may be an underestimate due to the low abundance of a particular target mRNA or an overestimate because 3'UTRs of different length can be overlapping. Thus, the best estimate is the total number of different miRNAs theoretically bound since this represents the level of possible regulation by miRNA in different cell types during different developmental stages (shaded columns in Tables 2-6). To some extent the number of sites is proportional to 3' UTR length (except for very short ones; Tables 2-4), but not proportional to length for the control genes (Table 6).

The top 20 binders of miRNA in the microRNA.org human genome are ranked according to the predicted numbers of different miRNAs bound listed in Table 2. *QKI* falls in the middle of this group at position 11. The top 20 miRNA binders all have long 3'UTRs (5,390 to 12,567 bp) and are potential targets of an impressive 51 to 69% of the total 677 defined human miRNAs in the microRNA database. This suggests a mechanism providing extensive and precise post-transcriptional control of gene expression in multiple cell types.

Next we looked at theoretical numbers of miRNAs bound in the other four members of the mammalian STAR family (Table 3). *QKI* tops the list of the other STAR members by an impressive 3 to 6.5 fold. Thus, high binding site number is not characteristic of the STAR family in general but rather unique to *QKI*. To examine whether the KH domain containing genes or genes with a neurological phenotype have a high number of miRNA binding sites, thirty genes in each of these categories were chosen at random and compared using the same criteria (Tables 4 and 5). When the total numbers of different miRNAs bound were analyzed, only one in each category; *AF4/FMR2 family, member 2* and *neurotrophic tyrosine kinase receptor, Type 2* (*NTRK2*) fell into the top 20 binders. Thus, in general neither KH domain encoding genes nor genes with neurological phenotypes have high numbers of theoretical miRNA binding sites. We performed a similar search using genes

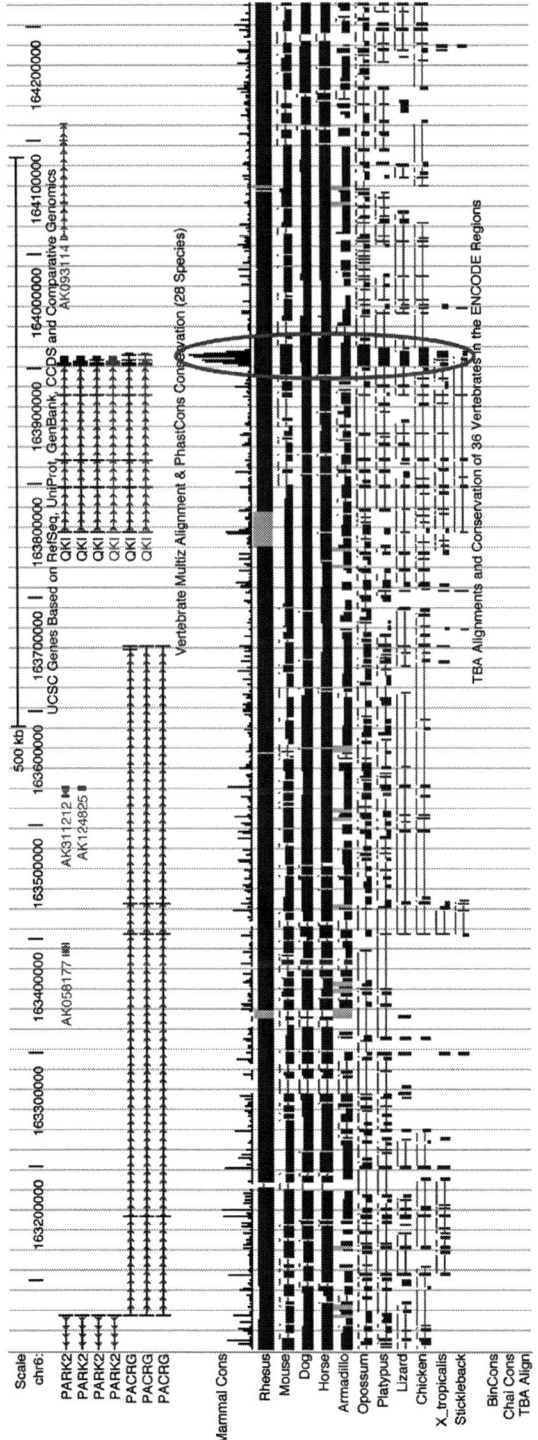

Figure 4. Extreme conservation in the *Qk* 3'UTRs. Conservation tracks taken directly from the UCSC human genome browser (hgs18) for 28 vertebrate species. The conservation of the two *Qk* 3'UTRs is circled. A color version of this image is available at www.landesbioscience.com/curie.

Table 2. Top 20 theoretical binders of miRNA in the human genome

No.	Gene Symbol	Gene Name	3'UTR Length	Ratio 3'UTR Length/ No. of Its Unique Sites	No. of Different miRNA Binding Sites	% Total Different Human miRNAs Bound (hu Total = 677)
1	*NRXN3*	Neurexin 3	6,981	15	470	69.4
2	*TNRC6B*	Trinucleotide repeat containing 6B	12,567	29	431	63.7
3	*SLC1A2*	Solute carrier family 1 (glial high affinity glutamate transporter), member 2	9,689	23	425	62.8
4	*MECP2*	Methyl CpG binding protein 2 (Rett syndrome)	8,554	20	419	61.9
5	*KLF12*	Kruppel-like factor 12	9,478	23	411	60.7
6	*PTPRT*	Protein tyrosine phosphatase, receptor type, T	7,717	19	410	60.6
7	*FBXW2*	F-box and WD repeat domain containing 2	7,583	19	398	58.8
8	*NTRK2*	Neurotrophic tyrosine kinase, receptor, Type 2	6,801	17	394	58.2
9	*NFAT5*	Nuclear factor of activated T-cells 5, tonicity-responsive	9,565	24	391	57.8
10	*NFIB*	Nuclear factor I/B.	6,461	17	389	57.5
11	**QKI**	**Quaking**	**6,399**	**17**	**386**	**57.0**
12	*FMR2, AFF2*	AF4/FMR2 family, member 2	9,331	24	385	56.9
13	*PURB*	Purine-rich element binding protein B	8,117	22	370	54.7
14	*KIAA2022*	KIAA2022	5,828	16	359	53.0
15	*IKZF2*	IKAROS family zinc finger 2	7,673	21	358	52.9
16	*TUG1*	Taurine upregulated gene 1	7,160	20	358	52.9
17	*C1orf21, DEN*	DENN/MADD domain containing 1B	9,465	27	357	52.7
18	*IGF2BP1*	Insulin-like growth factor 2 mRNA binding protein 1	6,701	19	350	51.7
19	*JHDM1D*	Jumonji C domain containing histone demethylase 1 homolog D (S. cerevisiae)	6,348	18	348	51.4
20	*KCNA1*	Potassium voltage-gated channel, shaker-related subfamily, member 1	5,390	15	348	51.4
				Average	388	

Table 3. The mammalian STAR family members

No.	Gene Symbol	Gene Name	3'UTR Length	Ratio 3'UTR Length/ No. of Its Unique Sites	No. of Different miRNA Binding Sites	% Total Different Human miRNAs Bound (hu Total = 677)
1	*QKI*	**Quaking**	**6,399**	17	**386**	57.0
2	*SF1*	Splicing factor 1	848	7	124	18.3
3	*Sam68, KHDRBS1*	KH domain containing, RNA binding, signal transduction associated 1	1,247	14	91	13.4
4	*SLM1, KHDRBS2*	KH domain containing, RNA binding, signal transduction associated 2	1,003	13	75	11.1
5	*SLM2, T-star, KH-DRBS3*	KH domain containing, RNA binding, signal transduction associated 3	521	9	59	8.7
				Average	147	

known to play a role in myelination and obtained the result that miRNAs bound were in the normal range (data not shown). To find an average total for randomly chosen cDNAs, we analyzed two more control groups of 30 KIAA and 30 FJL genes (Tables 6a and b). The average number of different miRNA sites was 97 and 64 respectively. Compared to 388, the average number of the top 20 genes listed in Table 2, randomly picked genes are several fold less.

To understand what the general norm is for a gene's potential number of miRNA binding sites, we performed a small-scale bioinformatics analysis. From the microRNA.org database, we analyzed all the 3,737 KIAA genes that contain at least one potential miRNA biding site. The data are plotted in Figure 5. The number of miRNA binding sites in the median cDNA 3'UTR is 54, 7 fold less than the group of top 20 genes (Table 2). In fact, the group of genes that contain more than 350 different miRNA sites numbers less than 40 cDNAs (about 0.1% of the human genome). Thus *QKI* belongs to this small group of genes that have the highest predicted number of miRNA binding sites in the genome.

DISCUSSION AND CONCLUSION

The STAR family appears to have an extremely diverse role in RNA metabolism. We have concentrated on QKI as an example of this pleiotropic activity and also presented some new data on the role of its conserved 3'UTRs gleaned from bioinformatics analysis.

The simplest way to explain the multiplicity of function of the STAR family is that individual members are turned on and off repetitively in many different tissues. This precise temporal and spatial distribution predicts a great many targets in different tissues as is indeed the case for *Qk*.[45] It is very likely that the *Qk* gene and its encoded proteins are kept under tight control with transcriptional, post-transcriptional and

Table 4. RNA processing genes encoding KH domains. A row for *Qk* has been inserted for comparison

No	Gene Symbol	Gene Name	3'UTR Length	Ratio 3'UTR Length/No. of Its Unique Sites	No. of Different miRNA Binding Sites
-	*QKI*	**Quaking**	**6,399**	**17**	**386**
1	**FMR2ª/AFF2**	**AF4/FMR2 family, member 2**	**9,331**	**24**	**383**
2	Igf2bp1	Insulin-like growth factor 2 mRNA binding protein 1	6,701	21	315
3	Nova1	Neuro-oncological ventral antigen 1	2,076	10	204
4	FMR1	Fragile X mental retardation 1	2,247	12	180
5	Igf2bp2	Insulin-like growth factor 2 mRNA binding protein 2	1,792	12	149
6	TDRKH	Tudor and KH domain containing	966	7	146
7	MEX3C/Rkhd2	Mex-3 homolog C (C. elegans)	1,801	13	142
8	Igf2bp3	Insulin-like growth factor 2 mRNA binding protein 3	2,153	16	135
9	MEX3B/Rkhd3	Mex-3 homolog B (C. elegans)	1,373	11	125
10	Fubp3	Far upstream element (FUSE) binding protein 3	1,330	11	121
11	AKAP1	A kinase (PRKA) anchor protein 1	2,444	21	117
12	HNRNPK	Heterogeneous nuclear ribonucleoprotein K	1,230	11	110
13	Fxr1	Fragile X mental retardation, autosomal homolog 1	1,027	9	109
14	Pcbp2	Poly(rC) binding protein 2	1,139	12	98
15	Ankrd17	Ankyrin repeat domain 17	1,446	15	95

continued on next page

Table 4. Continued

No	Gene Symbol	Gene Name	3'UTR Length	Ratio 3'UTR Length/No. of Its Unique Sites	No. of Different miRNA Binding Sites
16	Pcbp4	Poly(rC) binding protein 4	609	6	95
17	MEX3A/Rkhd4	Mex-3 homolog A (C. elegans)	2,283	26	88
18	FXR2	Fragile X mental retardation, autosomal homolog 2	616	8	79
19	PPP2R2C/PNO1	Protein phosphatase 2 (formerly 2A), regulatory subunit gamma isoform	2,724	37	74
20	Khsrp	KH-type splicing regulatory protein	764	17	45
21	Ankhd1	Ankyrin repeat and KH domain containing 1	438	11	40
22	Pcbp1	Poly(rC) binding protein 1	386	14	28
23	Nova2	Neuro-oncological ventral antigen 2	330	15	22
24	Bicc1	Bicaudal C homolog 1 (Drosophila)	194	10	19
25	MEX3D/Rkhd1	Mex-3 homolog D (C. elegans)	659	47	14
26	Hdlbp/vigilin	High density lipoprotein binding protein	879	98	9
27	KHDC1	KH homology domain containing 1	214	24	9
28	Rps3	Ribosomal protein S3	792	113	7
29	Dppa5	Developmental pluripotency associated 5	256	43	6
30	Pcbp3	Poly(rC) binding protein 3	747	747	1
				Average	99

Table 5. Genes chosen for neurological effects. A row for *Qk* has been inserted for comparison

No.	Gene Symbol	Gene Name	3'UTR Length	Ratio 3'UTR Length/No. of Its Unique Sites	No. of Different miRNA Binding Sites
1	**NTRK2**[a]	**Neurotrophic tyrosine kinase, receptor, Type 2**	**6,801**	**24**	**394**
-	**QKI**	**Quaking**	**6,399**	**13**	**386**
2	ATXN1	Ataxin 1	7,217	22	332
3	KCNMA1	Potassium large conductance calcium-activated channel, subfamily M, alpha member 1	2,389	7	323
4	DCX	Doublecortex; lissencephaly, X-linked (doublecortin)	7,908	25	320
5	ANK2	Ankyrin 2, neuronal	2,260	11	214
6	NEGR1	Neuronal growth regulator 1	4,504	24	185
7	NAB1	NGFI-A binding protein 1 (EGR1 binding protein 1)	2,433	14	174
8	BDNF	Brain-derived neurotrophic factor	2,926	19	153
9	PLP1	Proteolipid protein 1 (Pelizaeus–Merzbacher disease, spastic paraplegia 2, uncomplicated)	2,021	14	147
10	NF1	Neurofibromin 1 (neurofibromatosis, von Recklinghausen disease, Watson disease)	3,522	24	145
11	Robo1	Roundabout, axon guidance receptor, homolog 1 (Drosophila)	1,673	12	140
12	Ntn4	Netrin 4, Homo sapiens	1,279	10	132
13	GRIA2	Glutamate receptor, ionotropic, AMPA 2	2,642	20	130
14	NTRK3	Neurotrophic tyrosine kinase, receptor, Type 3	1,981	17	118
15	Reln	Reelin	1,022	10	100
16	MYT1	Myelin transcription factor 1	1,805	21	88

continued on next page

Table 5. Continued

No.	Gene Symbol	Gene Name	3'UTR Length	Ratio 3'UTR Length/No. of Its Unique Sites	No. of Different miRNA Binding Sites
17	Pmp22	Peripheral myelin protein 22	1,138	14	79
18	HAP1	Huntingtin-associated protein 1	2,062	32	64
19	als2	Amyotrophic lateral sclerosis 2 (juvenile)	1,297	22	60
20	NSG1	Neuron specific gene family member 1	1,623	27	60
21	Psen1	Presenilin 1 (Alzheimer's disease 3)	1,112	20	57
22	Smndc1	Survival motor neuron domain containing 1	1,110	19	57
23	MOG	MOG, myelin oligodendrocyte glycoprotein	1,187	25	47
24	LGI4	Leucine-rich repeat LGI family	1,494	36	42
25	HES1	Hairy and enhancer of split 1, (Drosophila)	374	11	34
26	AGR	Agrin	1,131	54	21
27	SMN2	Survival of motor neuron 2, centromeric	576	34	17
28	HTT	Huntingtin	3,901	279	14
29	Slit3	Slit homolog 3 (Drosophila)	179	18	10
30	Lrrn4	Eucine-rich repeats and calponin homology (CH) domain containing 4	754	126	6

[a]NTRK2 and FMR2 are also included in the 10 higest binders.

Table 6. Control KIAA and FLJ gene samples

No.	Gene Symbol	Gene Name	3'UTR Length	Ratio 3'UTR Length/ No. of Its Unique Sites	No. of Different miRNA Binding Sites
1	KIAA0351	RALGPS1, Ral GEF with PH domain and SH3 binding motif 1	4395	13	343
2	KIAA0410	NUPL1, nucleoporin like 1	4759	15	308
3	KIAA0314	AZ2A, bromodomain adjacent to zinc finger domain, 2A	3022	14	220
4	KIAA0429	MTSS1, metastasis suppressor 1	2202	12	186
5	KIAA0299	DOCK3, dedicator of cytokinesis 3	2639	16	162
6	KIAA0382	ARHGEF12, Rho guanine nucleotide exchange factor (GEF) 12	4859	30	162
7	KIAA0470	CEP170, centrosomal protein 170kDa	2021	14	146
8	KIAA0341	N4BP3, Nedd4 binding protein 3	4015	35	115
9	KIAA0457	DISC1, disrupted in schizophrenia 1	4441	40	112
10	KIAA0248	GBF1, golgi-specific brefeldin A resistance factor 1	556	5	110
11	KIAA0397	SGSM2, small G protein signaling modulator 2	1534	15	104
12	KIAA0078	RAD21, RAD21 homolog (S. pombe)	1565	16	99
13	KIAA0285	C2CD2L, C2CD2-like	861	9	96
14	KIAA0272	BAP1, BRCA1 associated protein-1 (ubiquitin carboxy-terminal hydrolase)	1279	16	79
15	KIAA0104	MRPL19, mitochondrial ribosomal protein L19	6924	92	75
16	KIAA0167	CENTG1, centaurin, gamma 1	1341	18	75

continued on next page

Table 6. Continued

No.	Gene Symbol	Gene Name	3'UTR Length	Ratio 3'UTR Length/ No. of Its Unique Sites	No. of Different miRNA Binding Sites
17	KIAA0442	AUTS2, autism susceptibility candidate 2	1872	25	75
18	KIAA0066	RAB3GAP1, RAB3 GTPase activating protein subunit 1 (catalytic)	1195	17	71
19	KIAA0193	SCRN1, secernin 1	3845	62	62
20	KIAA0039	POLD3, polymerase (DNA-directed), delta 3, accessory subunit	1964	32	61
21	KIAA0234	JARID1D, jumonji, AT rich interactive domain 1D	580	10	56
22	KIAA0026	MORF4L2, mortality factor 4 like 2	656	12	53
23	KIAA0091	MBTPS1, membrane-bound transcription factor peptidase, site 1	685	23	30
24	KIAA0054	HELZ, helicase with zinc finger	302	10	29
25	KIAA0218	TATDN2, TatD DNase domain containing 2	2036	85	24
26	KIAA0013	ARHGAP11A, Rho GTPase activating protein 11A	1822	101	18
27	KIAA0208	DVL3, dishevelled, dsh homolog 3 (Drosophila)	2771	198	14
28	KIAA0145	DGKD, diacylglycerol kinase, delta 130kDa	2637	264	10
29	KIAA0179	RRP1B, ribosomal RNA processing 1 homolog B (S. cerevisiae)	2696	270	10
30	KIAA0117	RBM34, RNA binding motif protein 34	79	11	7
		Average			97

continued on next page

Table 6. Continued

No.	Gene Symbol	Gene Name	3'UTR Length	Ratio 3'UTR Length/ No. of Its Unique Sites	No. of Different miRNA Binding Sites
1	FLJ21460	ATXN1L, ataxin 1-like	4402	21	211
2	FLJ10601	ANTXR1, anthrax toxin receptor 1	3685	20	188
3	FLJ20392	C14orf101, chromosome 14 open reading frame 101	2017	12	167
4	FLJ14442	AGBL4, ATP/GTP binding protein-like 4	1318	9	148
5	FLJ12015	BAALC, brain and acute leukemia, cytoplasmic	2312	16	148
6	FLJ12816	ABHD4, abhydrolase domain containing 4	1411	14	104
7	FLJ16146	CENTG3, centaurin, gamma 3	851	8	104
8	FLJ12666	C1orf108, chromosome 1 open reading frame 108	1252	12	102
9	FLJ39242	ACSF3, acyl-CoA synthetase family member 3	3365	36	94
10	FLJ37567	ARHGAP17, Rho GTPase activating protein 17	761	9	87
11	FLJ20093	ANKRD10, ankyrin repeat domain 10	1097	14	77
12	FLJ32820	ADAMTS14	1586	21	75
13	FLJ23751	ACPL2, acid phosphatase-like 2	1439	20	73
14	FLJ40094	ALPL, alkaline phosphatase, liver/bone/kidney	764	16	49
15	FLJ33340	C1orf201, chromosome 1 open reading frame 201	949	20	48
16	FLJ38678	C11orf74, chromosome 11 open reading frame 74	2292	50	46
17	FLJ11286	hypothetical protein FLJ11286	940	29	32
18	FLJ31001	AKNA, AT-hook transcription factor	891	29	31

continued on next page

Table 6. Continued

No.	Gene Symbol	Gene Name	3'UTR Length	Ratio 3'UTR Length/No. of Its Unique Sites	No. of Different miRNA Binding Sites
19	FLJ10342	Chromosome 6 open reading frame 166	331	14	23
20	FLJ32366	AP4S1, adaptor-related protein complex 4, sigma 1 subunit	515	32	16
21	FLJ23533	ACAD9, acyl-Coenzyme A dehydrogenase family, member 9	1950	139	14
22	FLJ14751	ALG10, asparagine-linked glycosylation 10 homolog (yeast, alpha-1,2-glucosyltransferase)	1386	107	13
23	FLJ11808	C7orf10, chromosome 7 open reading frame 10	279	23	12
24	FLJ10079	PDPR, pyruvate dehydrogenase phosphatase regulatory subunit	4402	367	12
25	FLJ25402	C8orf48, chromosome 8 open reading frame 48	336	34	10
26	FLJ22595	ARL14, ADP-ribosylation factor-like 14	522	58	9
27	FLJ21558	ASAH1, N-acylsphingosine amidohydrolase (acid ceramidase) 1	1109	139	8
28	FLJ10913	ADI1, acireductone dioxygenase 1	1043	149	7
29	FLJ36445	C19orf46, chromosome 19 open reading frame 46	237	59	4
30	FLJ21040	C16orf62, chromosome 16 open reading frame 62	686	229	3
				Average	64

In order to have a random selection of human genes we could not use the microRNA.org database because the queries are always returned ordered by the number of miRNA binding sites. Instead we searched MGI (http://www.informatics.jax.org/) with the term "KIAA". This returned a list of over 1000 mouse genes that had KIAA in their entry because it was or had been their human synonym. The list was ordered numerically and every tenth gene was chosen until 30 live entries were found in microRNA.org. Table 6b was done with the same procedure using "FLJ".

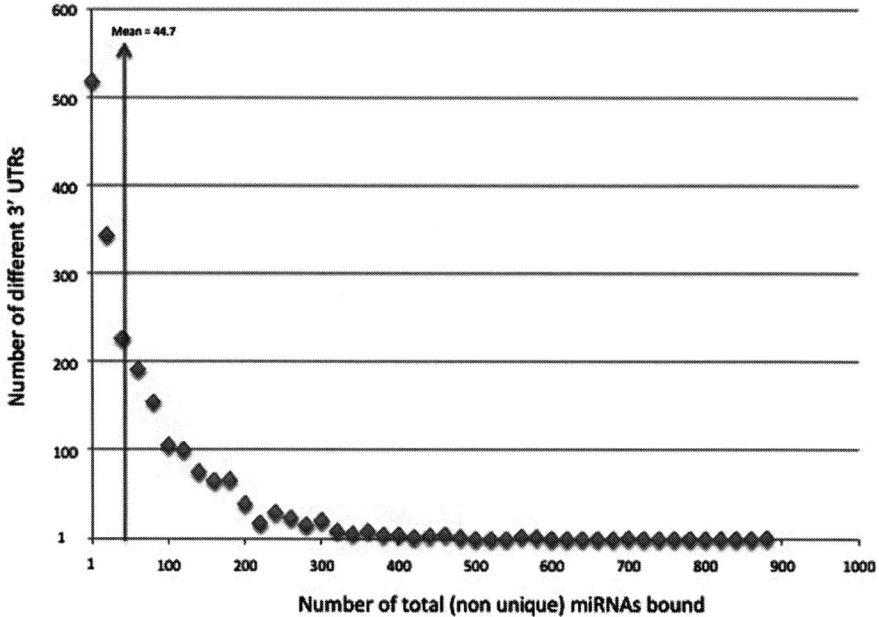

Figure 5. A scatter plot of the numbers of different 3'UTRs versus the theoretical miRNA binding sites. A sample of 3,737 human 'KIAA' gene 3'UTRs from microRNA.org database was analyzed. *QKI* does not appear in this sample because the mutation was named before the KIAA (Human cDNA project at the Kazusa DNA Research Institute) numbering system. An arrow indicates the miRNA binding sites of the median cDNA 3'UTR. A color version of this image is available at www.landesbioscience.com/curie.

post-translational mechanisms. With respect to transcription, there are possibly several alternate tissue specific promoters as demonstrated by the original viable mutation that contains a deletion of part of the upstream 5' regulatory region. The deletion leads to the impaired expression of QKI6 and QKI7 messages specifically in myelinating oligodentrocytes and Schwann cells, whereas QKI expression in other tissues are not affected.[28]

Another mechanism of fine-tuning is the balance of different splice forms since they may perform opposing functions. In the case of *Sam68*, its function could be regulated by the splice form that lacks the KH domain and therefore the ability to bind RNA.[74] Thus Sam68 heterodimers of KH-containing and non-KH-containing forms might tune down its RNA-binding activity. It appears that *Qk* also has such a splice form,[29] however its abundance is not known.

Notably, another level of potential tight control for *Qk* is mediated through its very long and conserved 3'UTRs. These 3'UTRs that theoretically bind an enormous variety of miRNAs (57% of the 677 defined human miRNAs) could provide fine-tuning of QKI at the mRNA level in different tissues during development. Some miRNAs that have frequent predicted binding sites in *Qk* 3'UTRs are specifically expressed in neural tissues,[75,76] myocardial, microvascular and endothelial tissues[77] and some immune cells.[78,79]

Thus the STAR family is like a conductor of an orchestra playing a fugue with peaks and valleys of voices in different tissues—It is possible that the STAR family members are master regulators of carrying external signals directly to RNA and thus are efficient at helping to maintain a dynamic and progressive equilibrium during development.

FUTURE APPLICATIONS, NEW RESEARCH, ANTICIPATED DEVELOPMENTS

With the recent progress of understanding STAR protein functions in signal transduction and RNA metabolism as well as their physiological roles in development and disease, the STAR family is emerging as key regulators of numerous biological processes. Further studies will be focused on addressing several important questions. (1) What are the RNA targets for STAR proteins? Besides predictions from the consensus sequence derived from in vitro experiments, the recently developed high throughput sequencing technology will enable the systematic identification of in vivo RNA targets of STAR proteins. (2) How are STAR family genes regulated transcriptionally by tissue specific enhancers and post-transcriptionally by splicing regulators and upstream miRNAs? (3) How are STAR proteins regulated by post-translational modifications and how these modifications regulate their molecular functions? (4) What are the signals upstream of all the STAR proteins? (5) What are the molecular mechanisms of STAR proteins linking external signals to RNA metabolism? And finally, (6) what are the biological functions of STAR proteins in different tissues and human diseases? Conditional mutations and mutations of multiple subfamily members will be needed to have a more comprehensive understanding of this question. The investigation of these questions has already started as is shown by the other chapters of this book. The STAR trek will continue and no doubt more interesting results will be revealed about this fascinating group of proteins.

ACKNOWLEDGEMENTS

This work was supported by the Endowed Scholar Fund from UT Southwestern Medical Center (J.I.W).

REFERENCES

1. Fumagalli S, Totty NF, Hsuan JJ et al. A target for src in mitosis. Nature 1994; 368:871-874.
2. Taylor SJ, Shalloway D. An RNA-binding protein associated with src through its SH2 and SH3 domains in mitosis. Nature 1994; 368:867-871.
3. Jones AR, Schedl T. Mutations in gld-1, a female germ cell-specific tumor suppressor gene in caenorhabditis elegans, affect a conserved domain also found in src-associated protein sam68. Genes Dev 1995; 9:1491-1504.
4. Ebersole TA, Chen Q, Justice MJ et al. The quaking gene product necessary in embryogenesis and myelination combines features of RNA binding and signal transduction proteins. Nat Genet 1996; 12:260-265.
5. Baehrecke EH. Who encodes a KH RNA binding protein that functions in muscle development. Development 1997; 124:1323-1332.
6. Zaffran S, Astier M, Gratecos D et al. The held out wings (how) drosophila gene encodes a putative RNA-binding protein involved in the control of muscular and cardiac activity. Development 1997; 124:2087-2098.
7. Arning S, Gruter P, Bilbe G et al. Mammalian splicing factor SF1 is encoded by variant cdnas and binds to RNA. RNA 1996; 2:794-810.

8. Di Fruscio M, Chen T, Richard S. Characterization of sam68-like mammalian proteins slm-1 and slm-2: Slm-1 is a src substrate during mitosis. Proc Natl Acad Sci USA 1999; 96:2710-2715.

9. Venables JP, Vernet C, Chew SL et al. T-star/etoile: A novel relative of sam68 that interacts with an RNA-binding protein implicated in spermatogenesis. Hum Mol Genet 1999; 8:959-969.

10. Ohno G, Hagiwara M, Kuroyanagi H. Star family RNA-binding protein asd-2 regulates developmental switching of mutually exclusive alteRNAtive splicing in vivo. Genes Dev 2008; 22:360-374.

11. Vega-Sanchez ME, Zeng L, Chen S et al. Spin1, a K homology domain protein negatively regulated and ubiquitinated by the E3 ubiquitin ligase SPL11, is involved in flowering time control in rice. Plant Cell 2008; 20:1456-1469.

12. Vernet C, Artzt K. Star, a gene family involved in signal transduction and activation of RNA. Trends Genet 1997; 13:479-484.

13. Lukong KE, Richard S. Sam68, the KH domain-containing superstar. Biochim Biophys Acta 2003; 1653:73-86.

14. Rajan P, Gaughan L, Dalgliesh C et al. Regulation of gene expression by the RNA-binding protein sam68 in cancer. Biochem Soc Trans 2008; 36:505-507.

15. Richard S, Yu D, Blumer KJ et al. Association of p62, a multifunctional SH2- and SH3-domain-binding protein, with src family tyrosine kinases, grb2 and phospholipase c gamma-1. Mol Cell Biol 1995; 15:186-197.

16. Najib S, Martin-Romero C, Gonzalez-Yanes C et al. Role of sam68 as an adaptor protein in signal transduction. Cell Mol Life Sci 2005; 62:36-43.

17. Matter N, Herrlich P, Konig H. Signal-dependent regulation of splicing via phosphorylation of sam68. Nature 2002; 420:691-695.

18. Cote J, Boisvert FM, Boulanger MC et al. Sam68 RNA binding protein is an in vivo substrate for protein arginine n-methyltransferase 1. Mol Biol Cell 2003; 14:274-287.

19. Babic I, Cherry E, Fujita DJ. SUMO modification of sam68 enhances its ability to repress cyclin d1 expression and inhibits its ability to induce apoptosis. Oncogene 2006; 25:4955-4964.

20. Siomi H, Dreyfuss G. RNA-binding proteins as regulators of gene expression. Curr Opin Genet Dev 1997; 7:345-353.

21. Zhang Y, Lu Z, Ku L et al. Tyrosine phosphorylation of QKI mediates developmental signals to regulate mRNA metabolism. EMBO J 2003; 22:1801-1810.

22. Lu Z, Ku L, Chen Y et al. Developmental abnormalities of myelin basic protein expression in fyn knock-out brain reveal a role of fyn in post-transcriptional regulation. J Biol Chem 2005; 280:389-395.

23. Liu Z, Luyten I, Bottomley MJ et al. Structural basis for recognition of the intron branch site RNA by splicing factor 1. Science 2001; 294:1098-1102.

24. Maguire ML, Guler-Gane G, Nietlispach D et al. Solution structure and backbone dynamics of the KH-QUA2 region of the xenopus star/gsg quaking protein. J Mol Biol 2005; 348:265-279.

25. Chen T, Richard S. Structure-function analysis of qk1: A lethal point mutation in mouse quaking prevents homodimerization. Mol Cell Biol 1998; 18:4863-4871.

26. Wu J, Zhou L, Tonissen K et al. The quaking I-5 protein (QKI-5) has a novel nuclear localization signal and shuttles between the nucleus and the cytoplasm. J Biol Chem 1999; 274:29202-29210.

27. Nabel-Rosen H, Dorevitch N, Reuveny A et al. The balance between two isoforms of the drosophila RNA-binding protein how controls tendon cell differentiation. Mol Cell 1999; 4:573-584.

28. Hardy RJ, Loushin CL, Friedrich VL Jr et al. Neural cell type-specific expression of QKI proteins is altered in quakingviable mutant mice. J Neurosci 1996; 16:7941-7949.

29. Kondo T, Furuta T, Mitsunaga K et al. Genomic organization and expression analysis of the mouse qkI locus. Mamm Genome 1999; 10:662-669.

30. Cox RD, Hugill A, Shedlovsky A et al. Contrasting effects of ENU induced embryonic lethal mutations of the quaking gene. Genomics 1999; 57:333-341.

31. Justice MJ, Bode VC. Three enu-induced alleles of the murine quaking locus are recessive embryonic lethal mutations. Genet Res 1988; 51:95-102.

32. Li Z, Takakura N, Oike Y et al. Defective smooth muscle development in qkI-deficient mice. Dev Growth Differ 2003; 45:449-462.

33. Lu Z, Zhang Y, Ku L et al. The quakingviable mutation affects qkI mRNA expression specifically in myelin-producing cells of the nervous system. Nucleic Acids Res 2003; 31:4616-4624.

34. Lorenzetti D, Antalffy B, Vogel H et al. The neurological mutant quaking(viable) is parkin deficient. Mamm Genome 2004; 15:210-217.

35. Noveroske JK, Lai L, Gaussin V et al. Quaking is essential for blood vessel development. Genesis 2002; 32:218-230.

36. Bohnsack BL, Lai L, Northrop JL et al. Visceral endoderm function is regulated by quaking and required for vascular development. Genesis 2006; 44:93-104.

37. Sidman RL, Dickie MM, Appel SH. Mutant mice (quaking and jimpy) with deficient myelination in the central nervous system. Science 1964; 144:309-311.

38. Noveroske JK, Hardy R, Dapper JD et al. A new ENU-induced allele of mouse quaking causes severe CNS dysmyelination. Mamm Genome 2005; 16:672-682.
39. Paronetto MP, Messina V, Bianchi E et al. Sam68 regulates translation of target mRNAs in male germ cells, necessary for mouse spermatogenesis. J Cell Biol 2009; 185:235-249.
40. Lukong KE, Richard S. Motor coordination defects in mice deficient for the sam68 RNA-binding protein. Behav Brain Res 2008; 189:357-363.
41. Richard S, Torabi N, Franco GV et al. Ablation of the sam68 RNA binding protein protects mice from age-related bone loss. PLoS Genet 2005; 1:e74.
42. Richard S, Vogel G, Huot ME et al. Sam68 haploinsufficiency delays onset of mammary tumorigenesis and metastasis. Oncogene 2008; 27:548-556.
43. Shitashige M, Naishiro Y, Idogawa M et al. Involvement of splicing factor-1 in beta-catenin/t-cell factor-4-mediated gene transactivation and pre-mRNA splicing. Gastroenterology 2007; 132:1039-1054.
44. Shitashige M, Satow R, Honda K et al. Increased susceptibility of sf1(+/–) mice to azoxymethane-induced colon tumorigenesis. Cancer Sci 2007; 98:1862-1867.
45. Galarneau A, Richard S. Target RNA motif and target mRNAs of the quaking STAR protein. Nat Struct Mol Biol 2005; 12:691-698.
46. Galarneau A, Richard S. The STAR RNA binding proteins GLD-1, QKI, SAM68 and SLM-2 bind bipartite RNA motifs. BMC Mol Biol 2009; 10:47.
47. Wu JI, Reed RB, Grabowski PJ et al. Function of quaking in myelination: Regulation of alternative splicing. Proc Natl Acad Sci USA 2002; 99:4233-4238.
48. Edenfeld G, Volohonsky G, Krukkert K et al. The splicing factor crooked neck associates with the RNA-binding protein HOW to control glial cell maturation in drosophila. Neuron 2006; 52:969-980.
49. Zhang Y, Feng Y. Distinct molecular mechanisms lead to diminished myelin basic protein and 2′,3′-cyclic nucleotide 3′-phosphodiesterase in qk(v) dysmyelination. J Neurochem 2001; 77:165-172.
50. Li Z, Zhang Y, Li D et al. Destabilization and mislocalization of myelin basic protein mRNAs in quaking dysmyelination lacking the QKI RNA-binding proteins. J Neurosci 2000; 20:4944-4953.
51. Larocque D, Pilotte J, Chen T et al. Nuclear retention of MBP mRNAs in the quaking viable mice. Neuron 2002; 36:815-829.
52. Larocque D, Galarneau A, Liu HN et al. Protection of p27(kip1) mRNA by quaking RNA binding proteins promotes oligodendrocyte differentiation. Nat Neurosci 2005; 8:27-33.
53. Larocque D, Fragoso G, Huang J et al. The QKI-6 and QKI-7 RNA binding proteins block proliferation and promote schwann cell myelination. PLoS One 2009; 4:e5867.
54. Zhao L, Ku L, Chen Y et al. Qki binds map1b mRNA and enhances MAP1b expression during oligodendrocyte development. Mol Biol Cell 2006; 17:4179-4186.
55. Jan E, Motzny CK, Graves LE et al. The STAR protein, GLD-1, is a translational regulator of sexual identity in caenorhabditis elegans. EMBO J 1999; 18:258-269.
56. Lakiza O, Frater L, Yoo Y et al. Star proteins quaking-6 and gld-1 regulate translation of the homologues gli1 and tra-1 through a conserved RNA 3′UTR-based mechanism. Dev Biol 2005; 287:98-110.
57. Saccomanno L, Loushin C, Jan E et al. The star protein QKI-6 is a translational repressor. Proc Natl Acad Sci USA 1999; 96:12605-12610.
58. Marin VA, Evans TC. Translational repression of a c. Elegans notch mRNA by the STAR/KH domain protein GLD-1. Development 2003; 130:2623-2632.
59. Mootz D, Ho DM, Hunter CP. The STAR/maxi-KH domain protein GLD-1 mediates a developmental switch in the translational control of c. Elegans pal-1. Development 2004; 131:3263-3272.
60. Schumacher B, Hanazawa M, Lee MH et al. Translational repression of c. Elegans p53 by gld-1 regulates DNA damage-induced apoptosis. Cell 2005; 120:357-368.
61. Hardy RJ. QKI expression is regulated during neuron-glial cell fate decisions. J Neurosci Res 1998; 54:46-57.
62. Lein ES, Hawrylycz MJ, Ao N et al. Genome-wide atlas of gene expression in the adult mouse brain. Nature 2007; 445:168-176.
63. Lim J, Hao T, Shaw C et al. A protein-protein interaction network for human inherited ataxias and disorders of purkinje cell degeneration. Cell 2006; 125:801-814.
64. Lee Y, Samaco RC, Gatchel JR et al. MiR-19, miR-101 and miR-130 co-regulate ATXN1 levels to potentially modulate SCA1 pathogenesis. Nat Neurosci 2008; 11:1137-1139.
65. Franklin RJ, Ffrench-Constant C. Remyelination in the CNS: From biology to therapy. Nat Rev Neurosci 2008; 9:839-855.
66. Bockbrader K, Feng Y. Essential function, sophisticated regulation and pathological impact of the selective RNA-binding protein QKI in CNS myelin development. Future Neurology 2008; 3:655-668.
67. Wu HY, Dawson MR, Reynolds R et al. Expression of QKI proteins and map1b identifies actively myelinating oligodendrocytes in adult rat brain. Mol Cell Neurosci 2001; 17:292-302.
67a. Keller A, Leidinger P, Lange J et al. Multiple Sclerosis: MicroRNA Expression Profiles Accurately Differentiate Patients with Relapsing-Remitting Disease from Healthy Controls. PLoS One. 2009; 4(10): e7440.

68. McInnes LA, Lauriat TL. RNA metabolism and dysmyelination in schizophrenia. Neurosci Biobehav Rev 2006; 30:551-561.
69. Aberg K, Saetre P, Jareborg N et al. Human QKI, a potential regulator of mRNA expression of human oligodendrocyte-related genes involved in schizophrenia. Proc Natl Acad Sci USA 2006; 103:7482-7487.
70. Aberg K, Saetre P, Lindholm E et al. Human QKI, a new candidate gene for schizophrenia involved in myelination. Am J Med Genet B Neuropsychiatr Genet 2006; 141B:84-90.
70a. Backx L, Fryns J-P, Marcelis C et al. Haploinsufficiency of the gene Quaking (QKI) is associated with the 6q terminal deletion syndrome. Am J Med Genet Part A 2010; 152A:319–326.
71. Zorn AM, Grow M, Patterson KD et al. Remarkable sequence conservation of transcripts encoding amphibian and mammalian homologues of quaking, a kh domain RNA-binding protein. Gene 1997; 188:199-206.
72. Betel D, Wilson M, Gabow A et al. The microRNA.Org resource: Targets and expression. Nucleic Acids Res 2008; 36:D149-153.
73. Consortium TGO. Gene ontology: Tool for the unification of biology. Nat Genet 2000; 25:25-29.
74. Barlat I, Maurier F, Duchesne M et al. A role for sam68 in cell cycle progression antagonized by a spliced variant within the kh domain. J Biol Chem 1997; 272:3129-3132.
75. Christensen M, Schratt GM. MicroRNA involvement in developmental and functional aspects of the nervous system and in neurological diseases. Neurosci Lett 2009.
76. Fiore R, Khudayberdiev S, Christensen M et al. Mef2-mediated transcription of the mir379-410 cluster regulates activity-dependent dendritogenesis by fine-tuning pumilio2 protein levels. EMBO J 2009; 28:697-710.
77. Urbich C, Kuehbacher A, Dimmeler S. Role of microRNAs in vascular diseases, inflammation and angiogenesis. Cardiovasc Res 2008; 79:581-588.
78. Lu LF, Liston A. MicroRNA in the immune system, microRNA as an immune system. Immunology 2009; 127:291-298.
79. Tsitsiou E, Lindsay MA. MicroRNAs and the immune response. Curr Opin Pharmacol 2009; 9:514-520.
80. Ruan J, Li H, Chen Z et al. Treefam: 2008 update. Nucleic Acids Res 2008; 36:D735-740.

CHAPTER 2

THE STAR FAMILY MEMBER
QKI and Cell Signaling

Yue Feng* and Andrew Bankston

Abstract: The family of Signal Transduction and Activators of RNA (STAR) is named based on the intriguing potential for these proteins to connect cell signaling directly to the homeostasis of their mRNA ligands. Besides the commonly shared single RNA binding domain that displays heterogeneous nuclear ribonucleoprotein K homology (KH), STAR proteins also harbor domains predicted to bind critical components in signal transduction pathways, in particular the Src-family protein tyrosine kinases (Src-PTKs). Indeed, accumulating evidence in recent years has demonstrated that the RNA-binding activity and the homeostasis of downstream mRNA targets of STAR proteins can be regulated by phosphorylation in response to various extracellular signals. This chapter provides a short review of the STAR member QKI, focusing on the essential role of QKI in development of the central nervous system, possible mechanisms by which QKI may link cell signaling to the cellular behavior of its mRNA targets and how QKI dysregulation may contribute to human diseases.

INTRODUCTION

The mouse *Qk* gene was cloned in 1996 by the Artzt group.[1] Soon after, the Xenopus *Qk* homologue[2] and Drosophila *Qk* homologue named held out wings(*How*)[3] were cloned, followed by the isolation of human *QKI* several years later.[4] The coding sequence and the genomic organization of the *Qk* gene are highly conserved in mammals.[4,5] The predicted features in QKI for binding RNA as well as signaling molecules and in Src-associated protein in mitosis 68 kDa (SAM68) prompted the idea that proteins in this family may function to connect cell signaling directly to mRNA homeostasis.[6] At least three major *Qk* isoforms are derived from alternative splicing of the 3′ coding exons.[1] Based on the

*Corresponding Author: Yue Feng—Emory University, 1510 Clifton Road, Atlanta, GA, 30322 USA.
 Email: yfeng@emory.edu

Post-Transcriptional Regulation by STAR Proteins: Control of RNA Metabolism in Development and Disease, edited by Talila Volk and Karen Artzt.
©2010 Landes Bioscience and Springer Science+Business Media.

length of *Qk* mRNAs, the corresponding QKI proteins are named QKI5, QKI6 and QKI7. A fourth isoform called QKI7b was identified later,[4,5] although expression of the encoded QKI7b protein has not been validated. All QKI proteins share the same N-terminus, which harbors an extended hnRNP K homology (KH) domain responsible for RNA-binding. Also present is a dimerization (QUA1) domain, an RNA binding stabilization domain (QUA2), several putative Src-homology 3 (SH3)-binding motifs that may interact with signaling factors, as well as a tyrosine cluster that serves as the phosphorylation sites for Src-PTKs.[7,8] The distinct C-termini of QKI isoforms determine their subcellular localization. The QKI5 C-terminus harbors a nuclear localization signal (NLS), resulting in predominant nuclear localization of QKI5 in the steady state, despite its ability for nuclear-cytoplasmic shuttling.[9] In contrast, QKI6 and QKI7 are mainly detected in the cytoplasm. Thus, although QKI isoforms harbor similar selectivity and affinity to the mRNA ligands in vitro,[8,10] the nuclear and cytoplasmic QKI isoforms are thought to exert differential and perhaps even opposing influence on the same mRNA ligands.

In vertebrates, QKI is essential for neural and endoderm development in embryos.[1,11] (The function of invertebrate QKI homologues is reviewed in other chapters.) However, the molecular and cellular mechanisms by which QKI controls development still remain poorly understood.[11] To date, the function of QKI is best characterized in the rodent nervous system,[7,12] taking advantage of the *quaking viable* (*Qk^v*) mutant mouse in which QKI expression is selectively attenuated in oligodendrocytes (OLs) and Schwann cells that are responsible for myelin formation in the central and peripheral nervous system respectively.[13-15] Extensive studies have demonstrated the essential role of QKI in controlling proliferation and differentiation of myelinating glial cells,[16,17] as well as in the actual ensheathment of axons by the specialized myelin membrane.[18-20] This is a key step in postnatal brain development that enables information to flow quickly between distant neurons. Consistent with these animal studies, accumulating evidence suggests the involvement of QKI hypofunction in white matter contributes to impairment in cognitive diseases represented by schizophrenia and depression.[21-24] In addition, emerging evidence points to the potential role of QKI in tumorigenesis in human glioma[4,25-27] and colon cancer.[28]

The biological function of QKI in development is carried out by controlling protein expression from its downstream mRNA targets via multiple post-transcriptional mechanisms.[7,29] Importantly, the RNA-binding activity of QKI is regulated by Src-PTK-dependent tyrosine phosphorylation.[8] Thus, delineating how QKI isoforms connect cell signaling to the cellular behavior of their mRNA targets will provide important clues in understanding fundamental rules that govern normal development, as well as the pathogenesis of human diseases caused by post-transcriptional dysregulation.

QKI IS ESSENTIAL FOR EMBRYONIC AND POSTNATAL DEVELOPMENT

In the developing brain, *Qk* gene products are initially detected in neural stem cells.[30] During neural cell fate decisions, *Qk* expression is selectively silenced in the neuronal lineage, but maintained in glia.[14,30] Thus, the QKI protein is postulated to govern neural cell fate specification,[30] although the functional requirement of QKI in this process has not been directly demonstrated. During postnatal development, the mRNAs encoding cytoplasmic QKI isoforms are gradually increased during accelerated myelin formation.[14] Meanwhile, the nuclear QKI5 declines.[14]—Change the vigorous alteration of many splicing factors during OL differentiation,[31] expression of QKI isoform proteins are conceivably regulated at the step of alternative splicing of *Qk* pre-mRNA. The reciprocal regulation of the nuclear and cytoplasmic QKI isoforms suggests their differential function. The first evidence that QKI proteins are indeed important for brain development is derived from the severe hypomyelination phenotype in the homozygous *quaking viable* (*Qk^v^/Qk^v^*) mutant.[32,33] *Qk^v^* is a spontaneous recessive mutation that leads to diminished QKI expression specifically in *Qk^v^/Qk^v^* OLs,[14] the myelinating glia responsible for the formation of myelin membrane on neuronal axons in the CNS. Consequently, many myelin structural protein mRNAs known to bind QKI are reduced in the *Qk^v^/Qk^v^* mutant brain.[18] Key among them is myelin basic protein (MBP), a component in compact myelin formation.[34] The *Qk^v^/Qk^v^* mutant only produces 5-10% of myelin as compared to that in the normal brain and such myelin fails to compact. Hence, the mutant develops severe tremors and seizures when myelin function becomes important at postnatal day 11~12.[12] Despite the fact that the *Qk^v^* lesion also deletes the *pacrg* and part of the *parkin* genes,[1] reintroducing expression of a *Qk*-transgene into OLs was sufficient to rescue the failure in myelination.[20] This clearly demonstrated the functional requirement of QKI in myelin development. More recent studies further reveal the essential role of QKI in controlling proliferation, differentiation and maturation of OL progenitor cells[16,17] and Schwann cells.[15] This in turn ensures a sufficient number of myelinating cells and the proper timing of myelination. Thus, QKI is a key factor that functions at multiple steps to advance OL and myelin development (more details are reviewed in ref. 7).

The *Qk* gene is widely expressed in various tissue types.[1,11] In addition, a consensus RNA motif identified for QKI binding in vitro, named QKI recognition element (QRE), is found in over 1000 mRNA species.[10] Many of these putative QKI targets encode proteins that play key roles in cell growth and differentiation, such as p27[kip1] that controls cell cycle exit.[17] Thus, QKI conceivably can exert profound influence on development, in addition to its well established role in the nervous system. Indeed, conventional knockout of the *QKI* gene in mice results in early embryonic lethality, due to severe developmental failure in many tissues including neural tube, cardiovascular system and smooth muscle.[11] More recent studies suggest that QKI not only plays important roles in controlling normal cell growth and development, but may also contribute to tumorigenesis in various cell types.[25,28] How QKI may mediate cell signaling to govern normal development and furthermore how QKI abnormality may be involved in pathogenesis of cancer and developmental diseases is an intriguing question that warrants rigorous investigation.

PHOSPHORYLATION OF QKI ISOFORMS BY Src-PTKS REGULATES THE CELLULAR FATE OF QKI mRNA TARGETS AT MULTIPLE POST-TRANSCRIPTIONAL STEPS

Several proline-rich SH3-binding motifs exist in all vertebrate QKI isoforms, leading to the hypothesis that QKI is a target of Src-PTKs,[6] whose RNA-binding activity may be governed by Src-PTK-dependent phosphorylation, similar to that of SAM68.[35] In addition, a cluster of five tyrosine residues is located immediately downstream of the SH3-binding motifs in QKI. Later studies demonstrated that the C-terminal tyrosine cluster, but not tyrosines in the KH domain or at the N-terminus, mediate phosphorylation of QKI by Src-PTKs in vitro, in transfected cells and in isolated myelin during brain development.[8] In addition, the predicted SH3-binding motifs are important for Src-dependent QKI phosphorylation.[8] However, unlike SAM68, QKI does not form a stable complex with Src-PTKs, suggesting a rather transient interaction between QKI and the kinase. Each QKI isoform is phosphorylated by Src-PTKs to a similar level in vitro and in transfected cells,[8] which negatively affects QKI binding to the MBP mRNA, the high affinity target of QKI in myelination.[20] However, whether Src-PTK-dependent phosphorylation represses QKI binding to all target mRNAs still remains to be determined.

QKI isoforms may exert a multitude of influence on their mRNA ligands at various post-transcriptional steps, including mRNA stability, translation, nucleo-cytoplasmic localization and alternative splicing.[29] The function of QKI in stabilizing mRNA targets is most extensively characterized in OL development in vivo as well as in cells transfected with reporter genes.[18,20,36] QKI deficiency causes destabilization of many mRNAs that encode key factors for OL differentiation and myelin synthesis.[18] However, mRNA stabilization does not necessarily enhance expression of the encoded protein, considering the differential nucleo-cytoplasmic distribution of the QKI isoforms. For example, due to the more severe reduction of cytoplasmic QKI6 and 7 rather than the nuclear QKI5 in the Qk^v/Qk^v mutant, the MBP mRNA is retained in the nucleus,[37] which contributes to the reduced MBP protein expression and myelin defects. In contrast, over-expression of the cytoplasmic QKI6 enhances MBP protein expression and rescues the myelination failure in the Qk^v/Qk^v mutant.[20] The differential biological function of nuclear and cytoplasmic QKI isoforms is also found in cell differentiation, in which only the cytoplasmic QKIs promote cell cycle exit while QKI5 appears to keep cells in proliferation status.[15,17] Thus, despite the fact that Src-PTK-dependent phosphorylation negatively affects QKI-RNA interaction,[8] opposing outcomes on QKI target protein expression can be mediated by different QKI isoforms. Presumably, Src-PTK-dependent phosphorylation of QKI5 may release nuclear retention of the mRNA ligands, thus enhancing the expression of the encoded proteins. Reciprocally, phosphorylation of cytoplasmic QKI isoforms may attenuate their influence to the cytoplasmic behavior of the mRNA targets.

How could the opposing effects on QKI-mRNA ligands by Src-PTK-dependent phosphorylation of QKI isoforms coordinately advance cell growth and development? It is important to point out that Src-PTK activity is also regulated during cell development.[38,39] Thus, the functional relationship between the developmental regulation of Src-PTK activity and QKI isoform expression and the distinct function of QKI target mRNAs at different developmental stages, are key factors for deciphering how QKI isoforms mediate Src-PTK signaling to govern normal development. The vigorous regulation of Src-PTK member Fyn, QKI isoforms and QKI target mRNAs in early OL differentiation as well as myelination offers an informative working model for connecting Src-PTK-QKI signaling to mRNA cellular behavior that ultimately advances myelination (Fig. 1).

Several Src-PTKs are expressed in OL progenitor cells. However, Fyn is the only Src-PTK member whose activity and expression are increased upon OL differentiation,[39] accompanied with a general down-regulation of the rest of the Src-PTKs.[40] Pharmacological inhibition and siRNA-mediated knockdown of Fyn attenuate OL differentiation,[39,41] indicating the essential role of Fyn in early OL development. However, upon the initiation of active myelin formation, Fyn activity markedly declines,[38] which is important for accelerated expression of myelin structural proteins such as MBP.[40] In mutant mice that lack Fyn activity, a significantly slower accumulation of MBP and hypomyelination are observed, regardless of the normal MBP expression at the early phase of myelination.

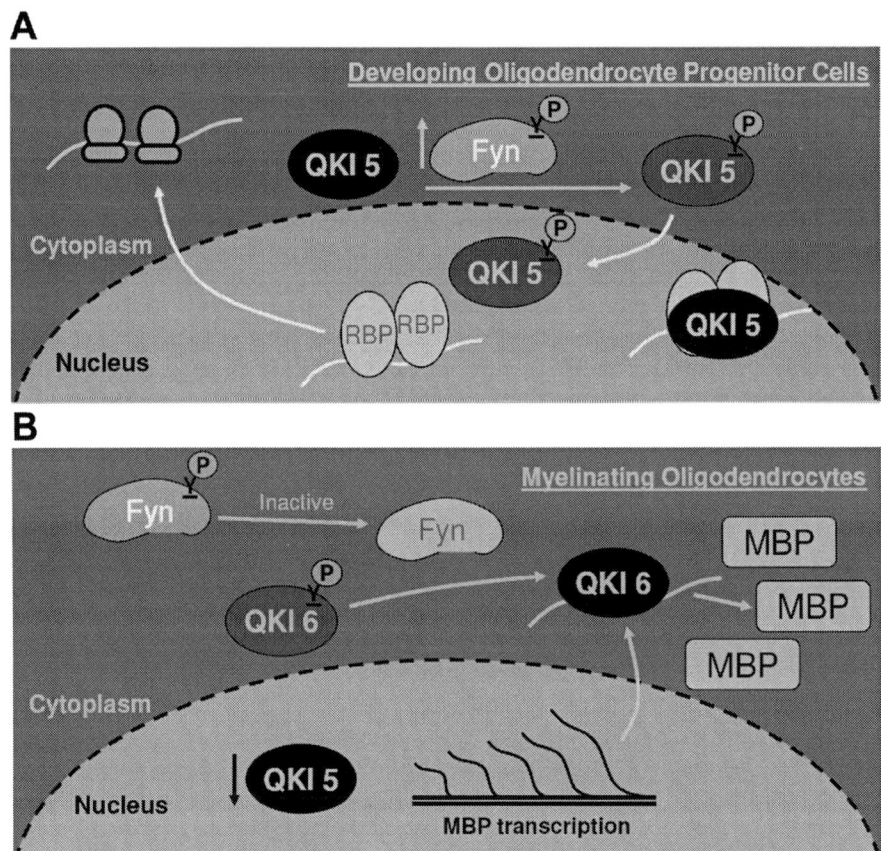

Figure 1. Connecting Fyn-signaling to mRNA cellular behavior by QKI isoforms. A) In early stage oligodendrocyte development, nuclear QKI5 is the predominant isoform. Unphosphorylated QKI5 (black) binds mRNA tightly and retains the mRNA targets in the nucleus, preventing differentiation. When signals trigger differentiation, Fyn activity is upregulated phosphorylating QKI5 at the C-terminal tyrosines, which decreases the RNA-binding affinity of QKI5, relieving nuclear retention of mRNA ligands from QKI5. These mRNAs, bound by other RNA-binding proteins (RBP), are then translocated to the cytoplasm for translation by ribosomes, which in turn support cell cycle exit and morphogenic differentiation. B) Upon oligodendrocyte maturation and active myelination, nuclear QKI5 is reduced concomitant with increased cytoplasmic QKIs, represented by QKI6. During this stage of development, transcription of myelin structural protein genes, represented by the MBP gene, is drastically upregulated and exceeds the capacity of nuclear retention by QKI5. Importantly, Fyn activity is markedly decreased, allowing the cytoplasmic QKIs to bind mRNAs, which is required for stabilization of QKI's target mRNAs. Y represents tyrosine.

During the early phase of OL progenitor cell development, the nuclear QKI5 is the predominant isoform,[14,16] that stabilizes the bound mRNA ligands. Known QKI targets include mRNAs that encode key proteins for controlling cell cycle progression and morphogenic differentiation, represented by p27[kip1] and the microtubule associated protein 1B (MAP1B), respectively.[17,36] Many of these QKI targets decline in later development, yet accurate timing for their transient expression is critical in governing proliferation, early differentiation and migration of OL progenitor cells. QKI targets that function later in myelination, such as the MBP mRNA, are either at low levels or not expressed. The vigorous increase of Fyn activity upon OL differentiation may help to release mRNAs from QKI5-dependent nuclear retention, which allows increased production of the corresponding proteins in the cytoplasm (Fig. 1A).

During OL maturation and active myelin formation, the nuclear QKI5 is markedly reduced in the brain, concomitant with increased cytoplasmic QKI isoforms.[14] This should attenuate the nuclear retention of QKI target mRNAs. At this stage, QKI target mRNAs encode many myelin structural proteins. Transcription of these myelin structural genes is markedly upregulated and a rapid accumulation of these QKI target mRNAs to exceptionally high levels is essential for accelerated myelin synthesis.[33] Despite the markedly enhanced transcription, these myelin structural mRNAs require protection by cytoplasmic QKI for survival from nuclease attack. QKI deficiency destabilizes these mRNAs, which in turn abrogates myelination in the Qk^v/Qk^v mice.[18] Expression of the cytoplasmic QKI6 isoform in OLs can rescue the aforementioned defects,[20] indicating the functional requirement of cytoplasmic QKI in later development. Importantly, Fyn activity markedly declines in the developing myelin.[38,40] Because Fyn negatively regulates QKI-MBP mRNA interaction,[8] the developmentally programmed down-regulation of Fyn offers a mechanism to increase the RNA-binding activity of QKI, which protects the MBP mRNA and accelerates the expression of myelin basic protein and myelin synthesis (Fig. 1B). Consistent with this idea, lack of the Fyn-QKI mediated acceleration mechanism leads to slow accumulation of the MBP mRNA, delayed myelin development and hypomyelination in both the fyn knockout mice and the Qk^v/Qk^v mutant,[18,40] with a more severe phenotype in the Qk^v/Qk^v mutant.

Besides stabilization of mRNA ligands, the cytoplasmic QKI6 isoform can also act as a translation suppressor.[42,43] In addition, the nuclear QKI5 is known to regulate alternative splicing of the myelin associated glycoprotein (MAG) pre-mRNA.[44] Mechanistically, QKI5 has been shown to bind a 53 nucleotide intronic sequence element and repress inclusion of the alternatively spliced exon in MAG.[44] Interestingly, this intronic target sequence for QKI5 is drastically different from the consensus QRE frequently found in the 3'UTR of QKI target mRNAs.[10] Whether and how Src-PTKs may regulate the activity of QKI in controlling splicing and/or translation still remains elusive.

NUMEROUS EXTRACELLULAR SIGNALS CAN BE LINKED TO THE Src-PTK-QKI PATHWAY

Src-PTKs represent the largest family of nonreceptor tyrosine kinases, which play crucial roles to control a diverse array of biological functions including proliferation, differentiation, cell shape, motility, migration, angiogenesis and survival.[45,46] In addition, Src-PTKs are known to govern the activity of neurotransmitter receptors[47] and signaling triggered by neuron-glia interaction.[38,48] Because the biology of Src-PTK-dependent QKI phosphorylation is best characterized in OL and myelin development with Fyn as the predominant Src-PTK member in OLs, we will focus on extracellular signals that are known to activate Fyn in OLs to discuss the function of QKI in cell signaling (Fig. 2).

Numerous extracellular signals directly or indirectly lead to Fyn activation in OLs, which in turn controls multiple aspects of OL and myelin development. Many growth factors required for OL progenitor cell survival trigger Fyn activation, such as the

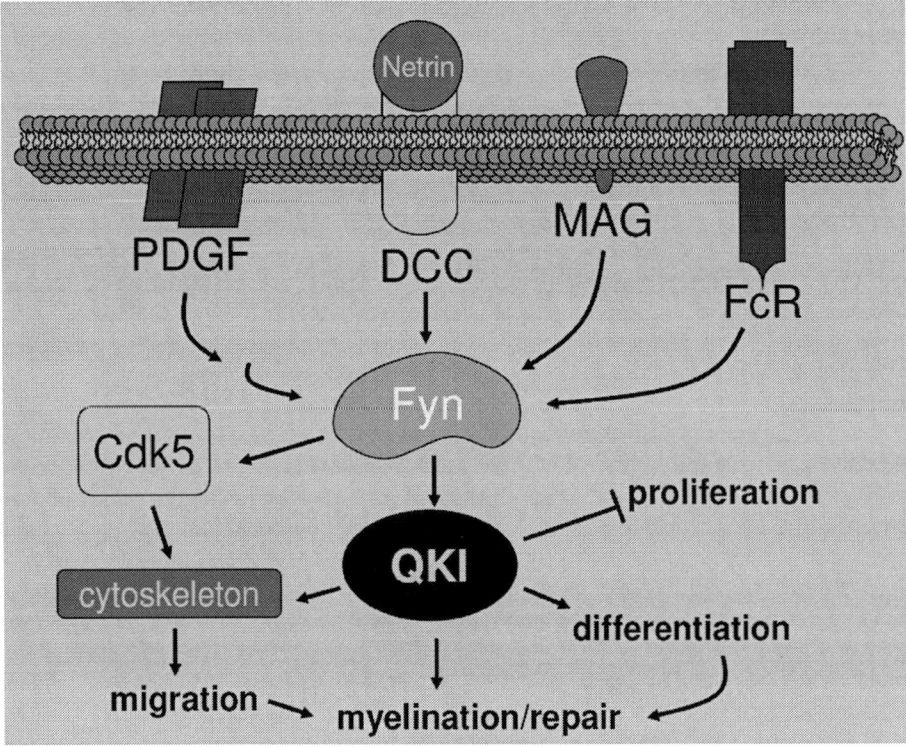

Figure 2. A hypothetical model describing the potential link of the Src-PTK-QKI pathway to extracellular signals throughout oligodendrocyte development. In oligodendroglia progenitor cells, platelet-derived growth factor (PDGF) binds its receptor PDGFR to trigger a signaling cascade that leads to Fyn activation, which in turn activates the serine/threonine kinase Cdk5 as well as QKI. Both of these can target microtubule associated protein 1B (MAP1B) for cytoskeleton re-organization and enhancing migration. As migrating oligodendrocytes approach the site of myelination, binding of Netrin 1 to its receptor, DCC, activates the Fyn-QKI pathway to stimulate oligodendrocyte processes extension and branching. Upon contact with the neuronal axon, myelination is initiated by MAG-dependent Fyn activation in mature oligodendrocytes. Finally, the FcR gamma/Fyn-QKI pathway may advance myelin repair after a demyelination event.

insulin-like growth factor-I (IGF-I) and the platelet-derived growth factor (PDGF).[49,50] In addition, PDGF is well characterized for its function in Fyn-dependent migration of OL progenitor cells.[44] During normal brain development as well as lesion repair, developing OLs migrate long distances to reach the site of myelin formation.[51] Upon binding to its receptor tyrosine kinase, PDGF triggers phosphorylation-dependent Fyn activation, which in turn activates the serine/threonine kinase CDK5.[50] CDK5 is capable of phosphorylating actin and microtubule associated proteins,[50,52] including the microtubule associated protein 1B (MAP1B),[52] a key factor that controls microtubule assembly and stabilization to drive early differentiation and migration of neurons and OLs.[36,53] Interestingly, our previous studies indicate that *MAP1B* mRNA is a target of QKI. Stabilization of *MAP1B* mRNA by QKI is an important mechanism for the vigorous upregulation of MAP1B in OL differentiation.[36] Presumably, Fyn-mediated QKI phosphorylation may contribute to MAP1B regulation. Thus, MAP1B may be a common target for PDGF-Fyn signaling via multiple mechanisms involving CDK5 and QKI.

The chemotropic guidance cue Netrin 1 has also been shown to activate Fyn. Netrin 1 binds its receptor, deleted in colorectal carcinoma (Dcc), on premyelinating as well as mature myelinating OLs,[54] which recruits Fyn to complex with the Dcc intracellular domain. Netrin1-Dcc signaling promotes OL process branching and myelin-like membrane sheath formation in a Fyn-dependent manner. OLs derived from *fyn knockout* mice failed to respond to Netrin-1-provoked process remodeling and myelin sheath formation.[54] Reorganization of both actin and microtubule cytoskeleton are critical for the extension and branching of OL processes. Thus, in addition to the known Netrin 1-Dcc targets RhoA and factors that control actin dynamics, the Fyn-QKI-MAP1B pathway should also contribute to OL morphogenesis (Fig. 2). In support of this idea, MAP1B is markedly upregulated, together with QKI7, upon initial contact between myelinating OLs and neuronal axons.

In fact, interaction between neuronal axons with OLs can also lead to Fyn activation directly. One example is the MAG-mediated Fyn activation in OLs upon OL-axon interaction in the initial phase of myelination.[38] As a member of the immunoglobulin superfamily, MAG binds the NOGO receptors on the neuronal axon surface, which functions in suppressing axonal growth and stabilization of axon-glia interactions.[55] Anti-MAG antibody-mediated cross-linking of MAG on the OL surface mimics axonal binding, which leads to hyperphosphorylation and activation of Fyn in OLs.[56] In addition, genetic abrogation of both MAG and Fyn drastically exacerbate the hypomyelination phenotype.[57] Besides MAG, the contactin complex coordinates signals from the extracellular matrix and the axonal surface to activate Fyn, which in turn regulates OL survival as well as myelination.[48] Furthermore, the gamma chain of immunoglobulin Fc receptors (FcR gamma) can activate Fyn.[58,59] The FcR gamma/Fyn signaling cascade is critically involved in OL differentiation and myelin repair,[58,59] suggesting a novel potential in the treatment of demyelinating diseases, represented by multiple sclerosis. Whether and how QKI functions as a downstream target in this signaling cascade to promote myelin repair is an intriguing possibility to be explored.

POTENTIAL ROLE OF QKI AND Src-PTK SIGNALING IN TUMORIGENESIS AND COGNITIVE DISEASES

A growing list of studies revealed that the *QKI* locus is frequently deleted in a subpopulation of human glioma.[25,26] In addition, in glioma samples in which the *QKI* gene is not deleted, diminished *QKI* isoform mRNAs are observed.[4] Interestingly, differential reduction of some but not all *QKI* isoforms are detected in glioma tumors, suggesting that post-transcriptional dysregulation of QKI may also occur. The role of QKI in promoting $p27^{kip1}$ expression and cell cycle exit,[17] together with the rich literature regarding the functional deficiency of $p27^{kip1}$ in glioma,[60,61] suggests that QKI deficiency likely contributes to $p27^{kip1}$ dysregulation in glioma. In addition to glioma, reduced QKI expression was also reported in human colon cancer, partly due to hypermethylation of the *QKI* promoter.[28] Moreover, forced expression of QKI blocks cell cycle progression and reduces the proliferation and tumorigenesis ability of colon epithelia. Thus, QKI potentially may function as a suppressor of tumorigenesis in various types of cancer.

On the other hand, Src-PTKs have been shown to underlie glioma-related proliferation, angiogenesis, migration and survival.[62] In addition, Src is frequently overexpressed and/or over-activated in human colorectal carcinoma,[63] which is known to associate with a poor clinical outcome.[62] The aberrantly increased Src activity in tumor cells likely attenuates the RNA-binding activity of QKI, which in turn can block the tumor suppressor function by QKI.

In addition to the possible involvement in tumorigenesis, hypofunction of QKI has recently been linked with the white matter impairment and myelin deficits in a number of cognitive disorders, including schizophrenics, major depression patients and suicide populations.[21,24] No deletion in the *QKI* gene promoter or coding region has been found in patients suffering with the aforementioned diseases. Instead, a single nucleotide polymorphism (SNP) in the intron up-stream of an alternatively spliced exon for *QKI5* was reported to segregate with probands in a large schizophrenia family.[22] In some reported cases, all *QKI* isoforms are reduced to similar levels in schizophrenia patient samples, likely due to epigenetic mechanisms that affect *QKI* transcription. However, in a large cohort, the cytoplasmic isoform *QKI7b* mRNA is preferentially reduced.[22] How antipsychotics and/or anti-depressants may influence *QKI* expression remains unknown. In addition, whether the aberrant *QKI* expression occurs in OLs or astroglia in patients who suffer from the aforementioned cognitive diseases is undetermined. Besides reduced *QKI* expression, SNPs in the *Fyn* gene and aberrant Fyn signaling are also found to be associated with schizophrenia.[64,65] Whether Fyn abnormalities may affect QKI function and contribute to the myelin impairment in schizophrenia remains to be elucidated.

CONCLUSION

Although the essential roles of QKI in advancing myelin development and Src-PTK-dependent tyrosine phosphorylation have been well established, whether QKI can be phosphorylated at serine/threonine by kinases in response to other signaling cascades still remains unknown. This is an intriguing possibility, especially considering the extensive phosphorylation of SAM68 on serine and threonine by various signaling mechanisms.[66,67] In addition, whether Src-PTKs modulate QKI function in a cell type- and mRNA ligand-specific manner are not understood. Despite the bioinformatic identification of putative QKI mRNA targets, among which many have been validated in normal brain development,[10] mRNA ligands of QKI in tumor cells have not been identified. Furthermore, post-transcriptional abnormalities in QKI targets, which in turn contribute to tumorigenesis and mental impairment, still remain unknown. These are prevailing issues in understanding how cell signaling is connected to mRNA cellular fate, that warrant vigorous investigation in the future.

ACKNOWLEDGEMENTS

We thank Danny Infield from Emory for critical editing of the text. This work is supported by NIH grant NS056097 and NMSS grant RG 4010-A-2 to YF. AB is supported by the NIH training grant T32 GM008367.

REFERENCES

1. Ebersole TA, Chen Q, Justice MJ et al. The quaking gene product necessary in embryogenesis and myelination combines features of RNA binding and signal transduction proteins [see comments]. Nat Genet 1996; 12:260-5.
2. Zorn AM, Krieg PA. The KH domain protein encoded by quaking functions as a dimer and is essential for notochord development in Xenopus embryos. Genes Dev 1997; 11:2176-90.
3. Zaffran S, Astier M, Gratecos D et al. The held out wings (how) Drosophila gene encodes a putative RNA-binding protein involved in the control of muscular and cardiac activity. Development 1997; 124:2087-98.
4. Li ZZ, Kondo T, Murata T et al. Expression of Hqk encoding a KH RNA binding protein is altered in human glioma. Jpn J Cancer Res 2002; 93:167-77.
5. Kondo T, Furuta T, Mitsunaga K et al. Genomic organization and expression analysis of the mouse qkI locus. Mammalian Genome 1999; 10:662-9.
6. Vernet C, Artzt K. STAR, a gene family involved in signal transduction and activation of RNA. Trends Genet 1997; 13:479-84.
7. Bockbrader K, Feng Y. Essential function, sophisticated regulation and pathological impact of the selective RNA-binding protein QKI in CNS myelin development. Future Neurol 2008; 3:655-68.
8. Zhang Y, Lu Z, Ku L et al. Tyrosine-phosphorylation of QKI mediates developmental signals to regulate mRNA metabolism. EMBO J 2003.
9. Wu J, Zhou L, Tonissen K et al. The quaking I-5 protein (QKI-5) has a novel nuclear localization signal and shuttles between the nucleus and the cytoplasm. J Biol Chem 1999; 274:29202-10.
10. Galarneau A, Richard S. Target RNA motif and target mRNAs of the Quaking STAR protein. Nat Struct Mol Biol 2005; 12:691-8.
11. Li Z, Takakura N, Oike Y et al. Defective smooth muscle development in qkI-deficient mice. Dev Growth Differ 2003; 45:449-62.
12. Hardy RJ. Molecular defects in the dysmyelinating mutant quaking. Journal of Neuroscience Research 1998; 51:417-22.
13. Lu Z, Zhang Y, Ku L et al. The quakingviable mutation affects qkI mRNA expression specifically in myelin-producing cells of the nervous system. Nucleic Acids Res 2003; 31:4616-24.
14. Hardy RJ, Loushin CL, Friedrich VL Jr et al. Neural cell type-specific expression of QKI proteins is altered in quakingviable mutant mice. Journal of Neuroscience 1996; 16:7941-9.
15. Larocque D, Fragoso G, Huang J et al. The QKI-6 and QKI-7 RNA binding proteins block proliferation and promote Schwann cell myelination. PLoS One 2009; 4:e5867.
16. Chen Y, Tian D, Ku L et al. The selective RNA-binding protein quaking I (QKI) is necessary and sufficient for promoting oligodendroglia differentiation. J Biol Chem 2007; 282:23553-60.
17. Larocque D, Galarneau A, Liu HN et al. Protection of p27(Kip1) mRNA by quaking RNA binding proteins promotes oligodendrocyte differentiation. Nat Neurosci 2005; 8:27-33.
18. Li Z, Zhang Y, Li D et al. Destabilization and mislocalization of myelin basic protein mRNAs in quaking dysmyelination lacking the QKI RNA-binding proteins. J Neurosci 2000; 20:4944-53.
19. Wu HY, Dawson MR, Reynolds R et al. Expression of QKI proteins and MAP1B identifies actively myelinating oligodendrocytes in adult rat brain. Mol Cell Neurosci 2001; 17:292-302.
20. Zhao L, Tian D, Xia M et al. Rescuing qkV dysmyelination by a single isoform of the selective RNA-binding protein QKI. J Neurosci 2006; 26:11278-86.
21. Aberg K, Saetre P, Lindholm E et al. Human QKI, a new candidate gene for schizophrenia involved in myelination. Am J Med Genet B Neuropsychiatr Genet 2006; 141:84-90.
22. Aberg K, Saetre P, Jareborg N et al. Human QKI, a potential regulator of mRNA expression of human oligodendrocyte-related genes involved in schizophrenia. Proc Natl Acad Sci USA 2006; 103:7482-7.
23. Haroutunian V, Katsel P, Dracheva S et al. The human homolog of the QKI gene affected in the severe dysmyelination "quaking" mouse phenotype: downregulated in multiple brain regions in schizophrenia. Am J Psychiatry 2006; 163:1834-7.
24. Klempan TA, Ernst C, Deleva V et al. Characterization of QKI Gene Expression, Genetics and Epigenetics in Suicide Victims with Major Depressive Disorder. Biol Psychiatry 2009.
25. Yin D, Ogawa S, Kawamata N et al. High-resolution genomic copy number profiling of glioblastoma multiforme by single nucleotide polymorphism DNA microarray. Mol Cancer Res 2009; 7:665-77.
26. Mulholland PJ, Fiegler H, Mazzanti C et al. Genomic profiling identifies discrete deletions associated with translocations in glioblastoma multiforme. Cell Cycle 2006; 5:783-91.
27. Ichimura K, Mungall AJ, Fiegler H et al. Small regions of overlapping deletions on 6q26 in human astrocytic tumours identified using chromosome 6 tile path array-CGH. Oncogene 2006; 25:1261-71.
28. Yang G, Fu H, Zhang J et al. RNA binding protein Quaking, a critical regulator of colon epithelial differentiation and a suppressor of colon cancer. Gastroenterology 2009.

29. McInnes LA, Lauriat TL. RNA metabolism and dysmyelination in schizophrenia. Neurosci Biobehav Rev 2006; 30:551-61.
30. Hardy RJ. QKI expression is regulated during neuron-glial cell fate decisions. J Neurosci Res 1998; 54:46-57.
31. Wang E, Dimova N, Cambi F. PLP/DM20 ratio is regulated by hnRNPH and F and a novel G-rich enhancer in oligodendrocytes. Nucleic Acids Res 2007; 35:4164-78.
32. Sidman S, Dickie M, Appel S. Mutant Mice (Quaking and Jimpy) with Deficient Myelination in the Central Nervous System. Science 1964; 144:309-11.
33. Campagnoni AT, Macklin WB. Cellular and molecular aspects of myelin protein gene expression. Mol Neurobiol 1988; 2:41-89.
34. Readhead C, Popko B, Takahashi N et al. Expression of a myelin basic protein gene in transgenic shiverer mice: correction of the dysmyelinating phenotype. Cell 1987; 48:703-12.
35. Paronetto MP, Achsel T, Massiello A et al. The RNA-binding protein Sam68 modulates the alternative splicing of Bcl-x. J Cell Biol 2007; 176:929-39.
36. Zhao L, Ku L, Chen Y et al. QKI binds MAP1B mRNA and enhances MAP1B expression during oligodendrocyte development. Mol Biol Cell 2006; 17:4179-86.
37. Larocque D, Pilotte J, Chen T et al. Nuclear retention of MBP mRNAs in the quaking viable mice. Neuron 2002; 36:815-29.
38. Umemori H, Sata S, Yagi T et al. Initial events of myelination involve Fyn tyrosine kinase signalling. Nature 1994; 367:572-6.
39. Osterhout DJ, Wolven A, Wolf RM et al. Morphological differentiation of oligodendrocytes requires activation of Fyn tyrosine kinase. J Cell Biol 1999; 145:1209-18.
40. Lu Z, Ku L, Chen Y et al. Developmental abnormalities of myelin basic protein expression in fyn knock-out brain reveal a role of Fyn in post-transcriptional regulation. J Biol Chem 2005; 280:389-95.
41. Colognato H, Ramachandrappa S, Olsen IM et al. Integrins direct Src family kinases to regulate distinct phases of oligodendrocyte development. J Cell Biol 2004; 167:365-75.
42. Saccomanno L, Loushin C, Jan E et al. The STAR protein QKI-6 is a translational repressor. Proc Natl Acad Sci USA 1999; 96:12605-10.
43. Lakiza O, Frater L, Yoo Y et al. STAR proteins quaking-6 and GLD-1 regulate translation of the homologues GLI1 and tra-1 through a conserved RNA 3'UTR-based mechanism. Dev Biol 2005; 287:98-110.
44. Wu JI, Reed RB, Grabowski PJ et al. Function of quaking in myelination: regulation of alternative splicing. Proc Natl Acad Sci USA 2002; 99:4233-8.
45. Frame MC. Src in cancer: deregulation and consequences for cell behaviour. Biochim Biophys Acta 2002; 1602:114-30.
46. Wheeler DL, Iida M, Dunn EF. The role of Src in solid tumors. Oncologist 2009; 14:667-78.
47. Kalia LV, Gingrich JR, Salter MW. Src in synaptic transmission and plasticity. Oncogene 2004; 23:8007-16.
48. Laursen LS, Chan CW, ffrench-Constant C. An integrin-contactin complex regulates CNS myelination by differential Fyn phosphorylation. J Neurosci 2009; 29:9174-85.
49. Cui QL, Zheng WH, Quirion R et al. Inhibition of Src-like kinases reveals Akt-dependent and -independent pathways in insulin-like growth factor I-mediated oligodendrocyte progenitor survival. J Biol Chem 2005; 280:8918-28.
50. Miyamoto Y, Yamauchi J, Tanoue A. Cdk5 phosphorylation of WAVE2 regulates oligodendrocyte precursor cell migration through nonreceptor tyrosine kinase Fyn. J Neurosci 2008; 28:8326-37.
51. Jarjour AA, Kennedy TE. Oligodendrocyte precursors on the move: mechanisms directing migration. Neuroscientist 2004; 10:99-105.
52. Hahn CM, Kleinholz H, Koester MP et al. Role of cyclin-dependent kinase 5 and its activator P35 in local axon and growth cone stabilization. Neuroscience 2005; 134:449-65.
53. Gonzalez-Billault C, Del Rio JA, Urena JM et al. A role of MAP1B in Reelin-dependent neuronal migration. Cereb Cortex 2005; 15:1134-45.
54. Rajasekharan S, Baker KA, Horn KE et al. Netrin 1 and Dcc regulate oligodendrocyte process branching and membrane extension via Fyn and RhoA. Development 2009; 136:415-26.
55. McKerracher L, Winton MJ. Nogo on the go. Neuron 2002; 36:345-8.
56. Marta CB, Taylor CM, Cheng S et al. Myelin associated glycoprotein cross-linking triggers its partitioning into lipid rafts, specific signaling events and cytoskeletal rearrangements in oligodendrocytes. Neuron Glia Biol 2004; 1:35-46.
57. Biffiger K, Bartsch S, Montag D et al. Severe hypomyelination of the murine CNS in the absence of myelin-associated glycoprotein and fyn tyrosine kinase. J Neurosci 2000; 20:7430-7.
58. Seiwa C, Yamamoto M, Tanaka K et al. Restoration of FcRgamma/Fyn signaling repairs central nervous system demyelination. J Neurosci Res 2007; 85:954-66.

59. Nakahara J, Tan-Takeuchi K, Seiwa C et al. Signaling via immunoglobulin Fc receptors induces oligodendrocyte precursor cell differentiation. Dev Cell 2003; 4:841-52.
60. Kirla RM, Haapasalo HK, Kalimo H et al. Low expression of p27 indicates a poor prognosis in patients with high-grade astrocytomas. Cancer 2003; 97:644-8.
61. Zagzag D, Blanco C, Friedlander DR et al. Expression of p27KIP1 in human gliomas: relationship between tumor grade, proliferation index and patient survival. Hum Pathol 2003; 34:48-53.
62. de Groot J, Milano V. Improving the prognosis for patients with glioblastoma: the rationale for targeting Src. J Neurooncol 2009.
63. Leroy C, Fialin C, Sirvent A et al. Quantitative phosphoproteomics reveals a cluster of tyrosine kinases that mediates SRC invasive activity in advanced colon carcinoma cells. Cancer Res 2009; 69:2279-86.
64. Hattori K, Fukuzako H, Hashiguchi T et al. Decreased expression of Fyn protein and disbalanced alternative splicing patterns in platelets from patients with schizophrenia. Psychiatry Res 2009; 168:119-28.
65. Rybakowski JK, Borkowska A, Skibinska M et al. Polymorphisms of the Fyn kinase gene and a performance on the Wisconsin Card Sorting Test in schizophrenia. Psychiatr Genet 2007; 17:201-4.
66. Paronetto MP, Zalfa F, Botti F et al. The nuclear RNA-binding protein Sam68 translocates to the cytoplasm and associates with the polysomes in mouse spermatocytes. Mol Biol Cell 2006; 17:14-24.
67. Resnick RJ, Taylor SJ, Lin Q et al. Phosphorylation of the Src substrate Sam68 by Cdc2 during mitosis. Oncogene 1997; 15:1247-53.

CHAPTER 3

INSIGHTS INTO THE STRUCTURAL BASIS OF RNA RECOGNITION BY STAR DOMAIN PROTEINS

Sean P. Ryder* and Francesca Massi

Abstract: STAR proteins regulate diverse cellular processes and control numerous developmental events. They function at the post-transcriptional level by regulating the stability, sub-cellular distribution, alternative splicing, or translational efficiency of specific mRNA targets. Significant effort has been expended to define the determinants of RNA recognition by STAR proteins, in hopes of identifying new mRNA targets that contribute their role in cellular metabolism and development. This work has lead to the extensive biochemical characterization of the nucleotide sequence specificity of a handful of STAR proteins. In contrast, little structural information is available to analyze the molecular basis of sequence specific RNA recognition by this protein family. This chapter reviews the relevant literature on STAR domain protein structure and provides insights into how these proteins discriminate between different RNA sequences.

INTRODUCTION

RNA-binding proteins play fundamental roles in cellular physiology. They guide decoding of the genome, they comprise the machinery that synthesizes proteins and they regulate the intensity, duration and sub-cellular distribution of gene expression.[1-6] Most RNA-binding proteins must distinguish between specific RNA sequences within the cellular milieu in order to function. The thermodynamic, kinetic and structural basis for sequence discrimination by RNA-binding proteins defines the mechanism of RNA recognition, which yields insights into specificity and biological function.

*Corresponding Author: Sean P. Ryder—Department of Biochemistry and Molecular Pharmacology, University of Massachusetts Medical School, 364 Plantation Street, LRB-906, Worcester, Massachusetts, 01605. Email: sean.ryder@umassmed.edu

Post-Transcriptional Regulation by STAR Proteins: Control of RNA Metabolism in Development and Disease, edited by Talila Volk and Karen Artzt.

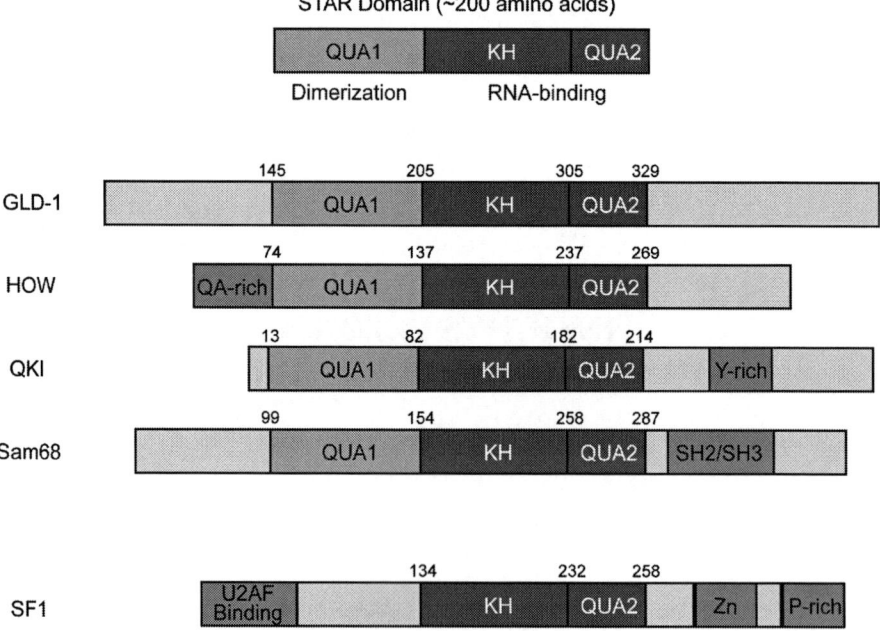

Figure 1. Domain structure of STAR domain protein examples GLD-1, HOW, QKI and Sam68. The domain structure of SF1 is shown for comparison. The QUA1 region, responsible for dimerization, and the KH and QUA2 regions, which form the RNA interface, are labeled. The approximate limits of each region are denoted. Other notable domains are presented in gray. A color version of this figure is available at www.landesbioscience.com/curie.

Sequence specific RNA-binding proteins employ one of three methods to discriminate between their RNA targets. Some recognize linear RNA sequences through base specific interactions, including hydrogen bonds and shape-specific van der Waals interactions.[7-10] Specificity arises through decoding the chemical differences between the four bases. Others recognize a specific three-dimensional structure, including stem-loops, pseudoknots, or helical junctions.[11-14] Here, specificity derives from the ability of a set of RNA sequences to fold into a recognizable shape. Lastly, some RNA-binding proteins bind to a protein-RNA complex and only achieve specificity through a combination of protein-protein and protein-RNA interactions.[15-17]

In this chapter, we focus on RNA binding by the STAR (signal transduction and activation of RNA) domain family of RNA-binding proteins (Fig. 1).[18] Genes that encode STAR proteins are found in the genomes of all metazoan species and have recently been identified in plants.[19] Highly studied examples include *Caenorhabditis elegans GLD-1, Drosophila melanogaster HOW* and their vertebrate homologs *Quaking (QKI)* and *Sam68*.[20-30] STAR proteins play a key role in developmental processes and have been implicated in human disease.[31,32] They couple cellular signaling events to post-transcriptional regulation of gene expression.

THE STAR DOMAIN

The STAR domain is a region of extended conservation flanking a canonical maxi-KH RNA binding domain (Fig. 1).[18] The region N-terminal to the maxi-KH domain is termed the QUA1 motif and the region C-terminal to the maxi-KH domain is termed the QUA2 motif. These regions are named after the mouse STAR domain protein Quaking (QKI) and define the difference between the STAR domain and other KH domain RNA-binding proteins.

Several lines of evidence indicate that the QUA1 domain is a homodimerization motif critical for STAR protein function.[33-35] First, in situ crosslinking studies in cell culture demonstrate that the STAR protein QKI self-associates.[33,34] Mutagenesis experiments reveal that self-association requires the QUA1 region. Moreover, immunoprecipitation experiments from mixed lysates of HeLa cells transfected with either myc-tagged or HA-tagged QKI show the two variants interact with each other and epitope-tagged variants are retained on a GST column when incubated with GST-QKI expressed in bacteria. A single point mutation within the QUA1 region of QKI eliminates dimerization both in vitro and in cell culture and causes an embryonic lethal phenotype in mice.[33] Subsequent experiments performed using recombinant variants of the *C. elegans* STAR protein (GLD-1) confirm this finding and demonstrate that the QUA1 domain is sufficient for stable dimerization in vitro.[36] Together, the data show that the QUA1 domain is both necessary and sufficient for dimerization and that dimerization is crucial for function in cells.

In contrast, the KH and QUA2 domains form an extended RNA-binding interface. Numerous qualitative and quantitative assays including UV crosslinking, column retention, gel mobility shift and fluorescence polarization (FP) experiments demonstrate that the KH and QUA2 domains bind to short penta- or hexanucleotide consensus sequences with moderate to high affinity.[27,36-42] The KH and QUA2 domains are sufficient for RNA-binding activity.[36] Dimerization improves affinity, likely mediated by direct interactions with RNA from both subunits of the dimer, although exactly how dimerization influences binding specificity remains unresolved. N-ethyl-N-nitrosourea (ENU) induced point mutations within the KH and QUA2 domains of QKI yield an embryonic lethal phenotype, demonstrating that both regions are required for function.[43] Similarly, mutations within the KH and QUA2 domain of GLD-1 have significant pleiotropic effects on germline development.[22] The data demonstrate that the activity of the RNA-binding subunits of STAR proteins is required for their biological function. In the next section, we review the quantitative data that defines the nucleotide sequence specificity of RNA recognition by GLD-1 and QKI, highlighting similarities and contrasting differences.

RNA RECOGNITION BY STAR PROTEINS

The nucleotide sequence specificity of four STAR domain proteins (GLD-1, QKI, HOW and Sam68) has been investigated in detail using quantitative in vitro methods. Two approaches have proven useful. In the first, the binding specificity is determined through identification of a regulatory target and characterization of the minimal binding sequence, followed by comprehensive mutagenesis.[36,44] In the second, the binding specificity is determined from a randomized sequence library using systematic evolution of ligands through exponential enrichment (SELEX) followed by computational comparison of the "winner" sequences to identify similarities.[27,38,45] Both methods yield comparable results, outlined below and indicate that GLD-1, QKI and HOW bind to RNA with similar though not identical specificity, while Sam68 binds to a different sequence.

Recognition of RNA by GLD-1

GLD-1 regulates the switch from spermatogenesis to oogenesis in *C. elegans* hermaphrodite development by controlling the expression of *tra-2*, a bifunctional membrane protein and transcription factor required for promoting oocyte cell fate.[42,46-48] Gel mobility shift and yeast 3-hybrid experiments reveal that GLD-1 binds to a repeat element in the 3'-UTR of the *tra-2* mRNA termed the TGE, for *tra-2* and *gli-1* element, with very high affinity ($K_{d, app} \sim 10$ nM) and a 2:1 apparent protein to RNA stoichiometry.[35,42] The TGE is 28 nucleotides in length and is primarily comprised of uridines and adenosines. Gel shift experiments with a battery of mutant variants across the TGE identify a bipartite recognition element that includes a UA dinucleotide near the 5'-end of the sequence and a contiguous UACUCA hexanucleotide element near the 3'-end of the sequence.[35] Because GLD-1 is a homodimer and because the recognition element is bipartite, it was proposed that one subunit recognizes the hexanucleotide element while the second subunit recognizes the UA dinucleotide as a partial or incomplete version of the hexanucleotide binding site.

Consistent with this hypothesis, a 12-nucleotide element comprised of the UACUCA hexanucleotide element flanked by three additional nucleotides on either side efficiently competes with the full-length TGE in competition gel shift experiments.[35] Moreover, isothermal titration calorimetry experiments demonstrate that the 12-nucleotide RNA binds to GLD-1 with an apparent 1:1 protein to RNA stoichiometry. Thus, each subunit of the dimer is capable of binding to an identical copy of the hexanucleotide sequence. Because there are three copies of the UACUCA hexanucleotide within the region of the *tra-2* UTR that contain the TGE repeats, it is not clear if the upstream UA dinucleotide is relevant to binding in worms, or if its apparent contribution to binding is an artifact of the minimal in vitro system.

To delineate the consensus GLD-1 binding sequence, a comprehensive library of single nucleotide mutations of the UACUCA sequence was analyzed within the context of the 12-nucleotide RNA.[35] Competition gel shifts were performed to determine the IC_{50} of the mutant sequence relative to the wild-type 12-mer RNA. The consensus recognition sequence, termed the STAR binding element (SBE), is 5'-UACU(C/A)A-3' (Table 1). Only the C to A mutation at the fifth position is tolerated without a reduction in competition efficiency. Allowing for mutations that reduce binding by up to 10-fold, a more relaxed consensus of 5'-(U>G>A/C)A(C>A)U(C/A>U)A-3' was also proposed. To date, the relative affinity and number of binding sites required for regulation have not been assessed in any functional assay in worms. Thus, it is not clear which consensus is more relevant to GLD-1 regulatory activity in worms, or if additional requirements beyond the determinants of binding in vitro are needed to select targets for regulation.

Recognition of RNA by QKI

QKI was identified as the gene responsible for the phenotype observed in the *Quaking* mouse (*Qk*[v]), a spontaneous mutant that arose in a mouse colony over forty years ago.[20] These mice fail to form compact myelin in their central nervous system, leading to a characteristic tremor upon movement. The *Qk*[v] allele is a large 1 MB deletion of chromosome 17 that modifies the expression pattern of the adjacent *Quaking* locus.[26,49,50] The dysmyelination

Table 1. Nucleotide sequence specificity of STAR domain proteins

Protein	Specificity	Method
GLD-1	5'-UACU(C/A)A-3' (conservative)	Gel Shift/
	5'(U>C>G/A)A(C>A)U(C/A)A-3' (relaxed)	Mutagenesis[36]
QKI[a]	5'-A(C/A)UAA-3'	FP/Mutagenesis[44]
QKI[b]	5'-ACUAA-3' (core)	
	5'-(U/C)AA(U/C)-3' (half-site)	SELEX[38]
HOW	5'-A(C>A)UAA-3'	Pull Down[37]
Sam68	5-UAAA-3'	SELEX[27]

[a] Binding specificity determined by Ryder and Williamson.[43]
[b] Binding specificity determined by Galarneau et al.[37]
RNA recognition by STAR domain proteins. The STAR protein identity is listed in the first column. The second column contains the RNA-binding consensus sequence. Degenerate nucleotides are contained within parentheses. The third column annotates the experimental method used to measure the specificity.

phenotype is due to the reduction of QKI expression in the oligodendrocyte lineage. There are three predominant isoforms of QKI, termed QKI5, QKI6 and QKI7 after their respective transcript lengths.[50] QKI5 is nuclear, while QKI6 and QKI7 are cytoplasmic.[51] All variants share the STAR domain and differ only in their C-terminus and 3'-UTR sequence. Several functions are proposed for QKI, including the regulation of mRNA stability, translation efficiency and alternative splicing.[41,52-55] Abundant evidence demonstrates that a major role of QKI in the oligodendrocyte lineage is in regulating the translation and stability of myelin basic protein (MBP) mRNA, which encodes a major myelin structural protein.[41,52] Reporter experiments and qualitative binding experiments indicate that QKI regulates MBP expression through specific association with it's 3'-UTR.[41]

Due to the high sequence similarity of QKI and GLD-1 within the STAR domain, it was predicted that these factors would bind to RNA with similar specificity. Indeed, QKI is capable of binding to TGE RNA in vitro and QKI-6 can functionally substitute for GLD-1 in a reporter assay in worms.[39,56] To directly measure the specificity of QKI, competition binding experiments were performed using an FP assay and the same 12-mer library used to map the GLD-1 consensus sequence.[39] The data reveal that QKI binds to RNA with similar, though not identical, specificity as GLD-1. The consensus sequence, termed the QKI STAR binding element (QSBE), is 5'-NA(A/C)UAA-3' (Table 1). This sequence is similar to the SBE recognized by GLD-1, but not identical. One difference is that the identity of the first nucleotide of the hexanucleotide consensus is not specified. A second difference is that an adenosine is permitted at the third position, where GLD-1 recognizes only a cytidine. The final and most dramatic difference is that an adenosine substitution at the fifth position of the consensus leads to a 40-fold enhancement of binding, whereas the same substitution in GLD-1 leads to a modest 2-fold increase. Several QKI consensus sites are present in the 3'-UTR of myelin basic protein transcripts.[39] QKI binds to all of these in vitro, but it binds with highest affinity (~10 nM) to a sequence present within a previously characterized region required for silencing MBP translation during localization.[57] The data suggest, but do not prove, that QKI regulates MBP mRNA through specific association with this element in its 3'-UTR. Functional studies are required to demonstrate that these binding sites are true *cis*-regulatory elements that confer QKI-dependent regulation of MBP expression in cells.

In an independent study, the binding specificity of QKI was determined using an in vitro SELEX protocol.[38] Several rounds of selection, enrichment and amplification lead to the identification of a number of aptamer sequences, many of which contained separate "core" 5'-NACUAAY-3' and "half site" 5'-YAAY-3' motifs with variable spacing (Table 1). Limited mutagenesis studies indicate that both elements are required for QKI to associate with the selected aptamers by gel mobility shift. There are two interesting differences between these results and the previous mutagenesis studies. First, the data suggest that an additional pyrimidine nucleotide is specified after the sixth position in the consensus. This requirement was not explicitly tested in previous binding experiments. Second, the selected consensus suggests a stricter requirement for a C at the third position, while the mutagenesis experiments indicate that a C or A is tolerated. Finally, the data suggest that QKI, like GLD-1, recognizes a bipartite element. It remains to be seen if either element is required for binding and function in vivo, or if both subunits of the QKI dimer can recognize either two copies of the "core" or two copies of the "half-site" element. Notably, the high affinity QKI binding site in the MBP 3'-UTR lacks a separate 5'-YAAY-3' half-site.[39] As with GLD-1, functional experiments are needed to identify the minimal requirements for QKI-dependent regulation in cells.

RNA Recognition by Other STAR Domain Proteins

A similar SELEX strategy was applied to Sam68, a STAR domain protein that is phosphorylated by Src and implicated in several aspects of cellular signaling and post-transcriptional regulation of gene expression.[24,25,27,58] The results suggest that Sam68 binds with high affinity to RNA sequences that contain the four nucleotide element 5'-UAAA-3' (Table 1).[27] Mutation of the element to 5'-UACA-3' eliminates binding in vitro, as does mutation of the STAR domain of the protein. Because reselection or comprehensive mutagenesis was not performed, the results do not comprehensively define the Sam68 consensus binding sequence. However, they clearly demonstrate that this protein binds to an element that is different than GLD-1 and QKI binding sites. The binding determinant is four nucleotides in length instead of five or six. Second, the site is more purine-rich than the sites recognized by the other proteins, including a run of three adenosine nucleotides. These results were recently confirmed in an independent selection and extended to another STAR protein Sam68-like mammalian protein 2 (SLM2, also known as Khdrbs3), which binds the same sequence.[45] SLM2 shows a greater degree of sequence similarity to Sam68 than to GLD-1, QKI, or other STAR proteins and as such is expected to bind RNA with specificity similar to Sam68.

The binding specificity of the *Drosophila* STAR domain protein HOW has also been investigated.[37] HOW is a post-transcriptional regulator of *stripe* and other transcripts required for wing tendon development.[30,40] The binding specificity of HOW was determined by mapping its binding sites in the 3'-UTR of *stripe* transcripts using biotinylated RNA fragments and streptavidin resin in HOW pull down assays.[37] HOW binds to RNA with almost identical specificity to QKI, recognizing the sequence 5'-NA(C>A)UAA-3' (Table 1). Intriguingly, HOW can bind to this sequence when it comprises the loop of a stem-loop structure, if the length of the loop is at least 12 nucleotides. Though not quantitative, an approximation of the relative affinity based upon relative signal intensity suggests that HOW binds more tightly to stem-loop RNA sequences than to unstructured RNAs. This result is in contrast to findings with QKI, where stem-loop sequences inhibit binding.[38] If RNA structure context is needed for HOW binding specificity in the functional context, then mRNA target specificity could arise from a combination of sequence and structure-based recognition, unique among the STAR domain proteins.

STAR DOMAIN STRUCTURE

What is the molecular basis for RNA-recognition by STAR domain proteins and what accounts for the differences in specificity between members of this family? To date, there are no structural data for an intact STAR domain from any protein. However, a pair of partial structures begins to reveal the three-dimensional architecture of this domain and the structural basis for sequence-specific RNA recognition.[8,59] The next few sections will review the structures, including an nuclear magnetic resonance (NMR) spectroscopy structure of the KH-QUA2 domain of Splicing Factor 1 (SF1, Fig. 1) and the NMR structure of the KH-QUA2 domain of QKI. A comparison of structure-based sequence alignments, homology modeling and protein mutagenesis experiments provide a starting point to understand how RNA binding specificity is achieved.

The NMR Structure of SF1 Bound to RNA

SF1 is a component of the eukaryotic splicing apparatus that is critical for recognition of the branch site adenosine in introns.[60-62] SF1 is the mammalian homolog of the yeast branch point binding protein (BBP), which recognizes the branch site adenosine within the context of the branch point sequence (BPS) 5'-UACUAAC-3'.[62] SF1 contains recognizable maxi-KH and QUA2 domain, but lacks the QUA1 domain typical of STAR proteins (Fig. 1).[8] Instead, it binds to RNA cooperatively with U2AF35/65, which recognizes the polypyrimidine tract and 3'-splice site to form the initial intron recognition complex.[17] Intriguingly, the BPS sequence lies within the binding consensus sequence recognized by GLD-1, QKI and HOW.[36-39] The structure of mammalian SF1 bound to the yeast BPS has been determined by NMR, providing the first glimpse into the structural basis of RNA recognition by the KH and QUA2 domains.[8] In the following section, we provide an overview of the NMR structure of SF1, outlining the amino acids that comprise the RNA-binding interface.

Overview of the Structure

The three dimensional structure of SF1 KH-QUA2 region bound to an 11-nucleotide RNA containing the yeast BPS (5'-UAUACUAACAA-3') has been determined using NMR spectroscopy.[8] The high quality of the experimental data, including the large number of intermolecular restraints, enables the high-resolution structural characterization of the SF1/RNA complex (Fig. 2A). The most interesting feature of the structure is the relative organization of the KH and QUA2 domains. QUA2 forms an alpha helix that packs against the maxi-KH domain, forming an expanded structure with topology β1-α1-α2-β2-β3-α3-α4. This organization forms an extended hydrophobic surface between the two domains that comprises part of the RNA binding groove. The RNA molecule is bound in an extended conformation onto a large RNA binding surface that includes helix α4 of QUA2 and the following elements of KH: helices α1 and α2, strand β2, the conserved GXXG loop (situated between α1 and α2) and the variable loop (also known as the "thumb" region, between β2 and β3) that encloses the RNA.

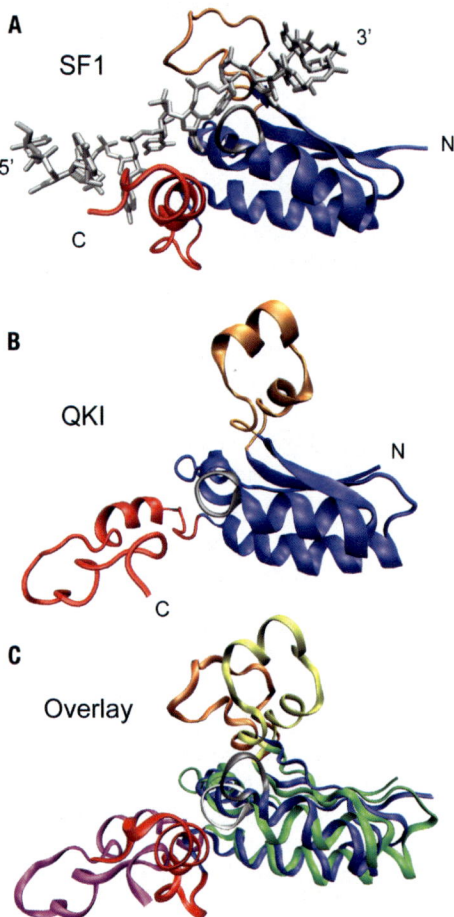

Figure 2. Structures of the KH and QUA2 regions. Each image represents a single model from within a family of structures deposited within the respective PDB files. A) NMR structure of *H. sapiens* SF1 KH and QUA2 region bound to BPS RNA. The 5′ and 3′ ends of the RNA are labeled, as is the N and C termini of the protein. The KH domain is in blue, the QUA2 region in red, the variable loop is colored orange and the GXXG loop is black. The RNA sequence is gray. B) NMR structure of the KH and QUA2 region of *X. laevis* QKI in the absence of RNA. The coloring scheme is the same as panel A. C) Overlay of the SF1 structure with the QKI structure, with RNA removed for clarity. The KH, QUA2, variable loop and GXXG loop of SF1 are identical to panel A. The KH domain of QKI is represented in green, the QUA2 region in magenta, the variable loop in yellow and the GXXG loop in white.

The QUA2 domain is involved in the recognition of the nucleotides at the 5′ end of the RNA (A4-C5), while the KH domain is critical for the recognition of the 3′ end of the BPS RNA (U6-A7-A8-C9). The SF-1/RNA complex is stabilized by a combination of extensive hydrophobic interactions with aliphatic side chains, base-specific and backbone hydrogen bonds, as well as electrostatic interactions. A peculiar feature of SF1, relative to other single-stranded RNA binding proteins, is the absence of aromatic-base stacking interactions between the protein and the RNA.[7,9]

Recognition of RNA by SF1

Analysis of the NMR secondary chemical shifts collected in the absence of RNA indicates that the overall structure of SF1 is largely maintained in the free state.[8] Thus, it is likely that SF1 associates with BPS RNA through a preformed RNA interface, rather than through large structural rearrangements induced by association of the RNA. This does not preclude local structural re-arrangements necessary to form the stable interactions observed in the structure calculations. The detailed interactions responsible for sequence-specific recognition of BPS RNA by SF1 are described below.

Residues in the QUA2 domain of SF1 recognize A4 and C5. In particular, A4 is recognized through hydrophobic interactions between the base and the side chains of Arg 255 and Ala 248. The shape of the pocket formed by these amino acids mediates specificity for an adenosine residue. The base and sugar of C5 form hydrophobic interactions with the side chains of Leu 244 and of Leu 247 of QUA2, respectively. Specificity for a cytidine is provided in the form of a hydrogen bond between the side chain amide of Asn 151 and the N5 of C5.

U6 binds to the interface between the QUA2 and KH domains. This position is stabilized through van der Waals interactions between the base and the side chains of Leu 244 and of Leu 247, located in the QUA2 domain, as well as by hydrogen bonds with the backbone of Leu 155 and Gly158 of the KH domain (Fig. 3A). The sugar of U6 is stabilized by a hydrogen bond with the backbone amide of Arg 160 and is stacked against the side chain of Pro 159, in the conserved GPRG loop. This nucleotide is packed against the backbone atoms of Gly 154, explaining the conservation of a glycine residue at this position. The steric constraints and the specific hydrogen bond interactions formed by the U6 recognition pocket between the KH and QUA2 domains of SF1 explain the strong selectivity of this position.

A7 interacts exclusively with residues located in the KH domain. The conformation of A7 in the complex is stabilized by a π-cation interaction with the sidechain of a highly conserved lysine (Lys 184) located in the variable loop (Fig 3B). In addition, a hydrogen bond between N6 of A7 and the sidechain of Glu 159 and van der Waals interaction with Val 153 contribute to the recognition of this base in the protein-RNA complex.

The branch point adenosine A8 is bound through specific hydrogen bonds with the protein backbone amide and carbonyl groups of Ile 177 (Fig. 3C). An additional hydrogen bond between N7 of A8 and the 2' hydroxyl of A7 contributes to the unique recognition of an adenine base at this position. A series of conserved aliphatic residues of the KH domain, Ile 157, Leu 164, Ile 175, Ile 177 and Val 183, form a hydrophobic envelope that surrounds A8. The close proximity of the phosphate groups of A7 and A8 to the backbone of the conserved GPRG loop is consistent with the abolished or severely diminished RNA binding affinity associated with mutations of these residues to negatively charged glutamate or aspartate.

C9 recognition is achieved through hydrophobic interactions of the pyrimidine base with the side chain of Val 183 and of the ribose ring with the side chain of Met 176. Compared to other positions, the identity of the C residue at this position is not well specified by the nature of the interactions. A8, C9 and A10 are stacked in an A-helix-like conformation, thus base stacking interactions stabilize the nucleotides in this conformation.

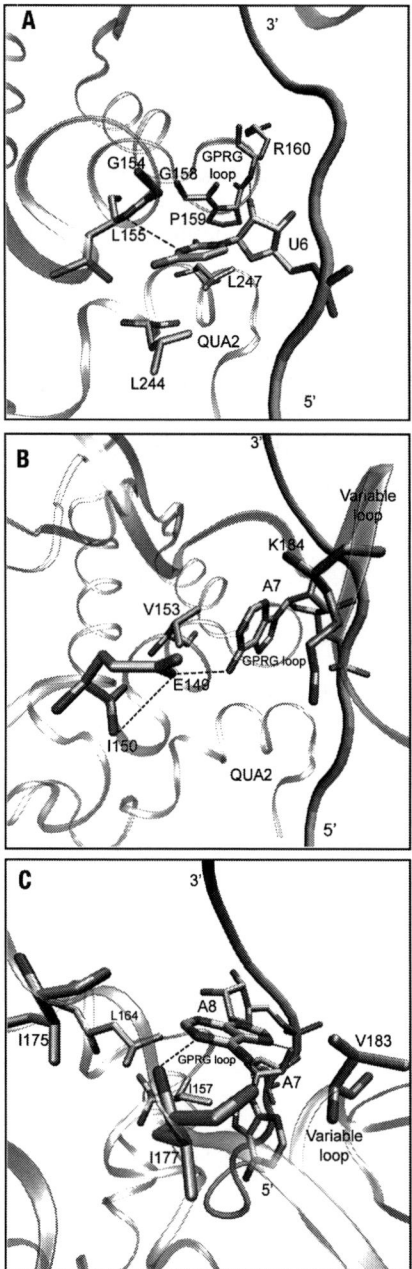

Figure 3. Close up view of sequence specific RNA interactions in the SF1-RNA complex NMR structure. A) The U6 binding pocket is shown. Amino acid side chains and the uridine nucleotide are labeled. A hydrogen bond between the backbone amide and the N3 of uridine is represented with a dashed line. B) The A7 binding pocket. The amino acid side chains involved in the interaction are labeled. A base-specific hydrogen bond between the N6 exocyclic amine of A7 and a Glu 149 is represented with a dashed line. C) The A8 binding pocket. Amino acids and base specific hydrogen bonds are labeled as in panels A and B.

The NMR Structure of the KH and QUA2 Regions of QKI

The structure of the KH-QUA2 region of the *Xenopus laevis* Quaking protein (XQUA) has been solved using NMR spectroscopic methods.[59] In contrast to the SF1 structure described above, the QKI structure was obtained in the absence of RNA. The ensemble of calculated structures indicate that the KH domain of the protein is well defined, with the exception of the variable loop which lacks long-range restraints. The QUA2 domain clearly presents an α-helical secondary structure, but its orientation relative to the KH domain is undefined as no long-range distance restraints between the two domains were identified. A representative structure of the ensemble is shown in Figure 2b. Apart from the difference in defined orientation between the QUA2 domains of SF1 and Xenopus QKI, a major variation between the two structures is between the variable loops, colored in orange. In QKI, this region is α-helical and more precisely organized into two α-helices. As a consequence, the KH domain is characterized by the following topology: β1-α1-α2-β2-α3-α4-β3-α5.

Characterization of the backbone dynamics of QKI through measurement of the ^{15}N relaxation rates demonstrates that the KH domain is well structured in solution in its entirety, including the variable loop region. Thus, the heterogeneity of the structures observed in this region is entirely due to the paucity of long-range distance restraints. The structure of this region is well defined internally, but its orientation relative to the rest of the KH domain is less clear. In agreement with the NMR studies of SF1 that indicate that the QUA2 region is α-helical in the absence of RNA, this study of QKI in the free state supports the presence of an α-helix (residues 189-201) in the QUA2 region.[8,59] [15]N relaxation studies indicate that the QUA2 domain of the protein is more dynamic than the KH domain. The heterogeneity of the ensemble of structures in the QUA2 domain is due to the presence of a highly flexible linker connecting the α-helix region of QUA2 and the KH domain. Hence, the QUA2 α-helix is not docked against the KH domain, as observed in the bound structure of SF1, but instead is mobile in solution.

Figure 2C shows a comparison of the QKI structure to SF-1/BPS RNA complex structure, where the RNA has been removed for clarity.[8,59] The major differences lie within the orientation of the GXXG loop and the presence of two α-helical elements in the variable loop. In models where the RNA from the SF1 structure is docked into the QKI structure without structural refinement, the topology of the QKI GXXG loop would cause steric clash with U6 and A7. Moreover, α4 within the QKI variable loop would clash with A11, although the differences in the amino acids interacting with this base and biochemical evidence that the nucleotide identity is not specified at this position in QKI suggest differences in RNA recognition at this position. It is likely that such dissimilarities between the structures of the two proteins arise from small differences between the conformations of the free and bound states. The highly dynamical character of QUA2 observed for QKI in the free state support a mechanism where the KH domain of the STAR domain proteins is the first to interact with the RNA, followed by the docking of the QUA2 domain helix against the KH domain. Kinetic studies are needed to tease apart the contribution of each region towards the rate of RNA binding.

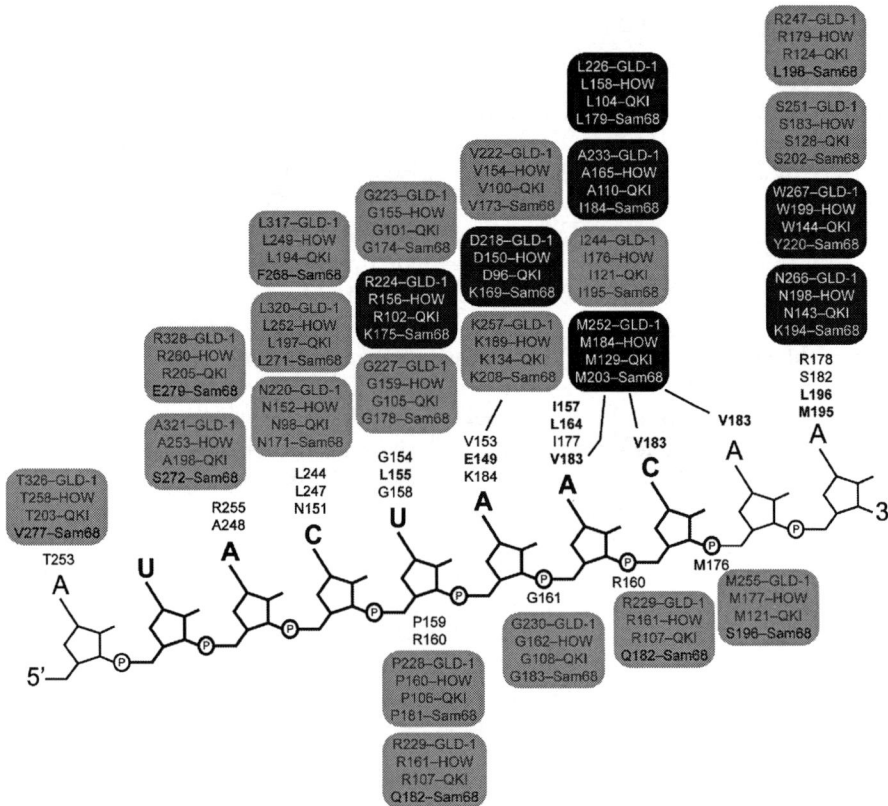

Figure 4. Diagram of the protein-RNA contacts observed in the SF1-BPS RNA complex structure. The sequence of the RNA is shown next to a backbone diagram. The UACUAA element is bold. The SF1 amino acids that contact RNA are presented next to the nucleotide they interact with in the structure. The corresponding amino acid in GLD-1, HOW, QKI and Sam68 is given in a box above. Conserved amino acids are in gray. Blue boxes represent positions where all four proteins differ from SF1, but the identity of the amino acid difference is the same in GLD-1, QKI and HOW. Red boxes represent positions where all four proteins differ from SF1 but the difference is identical in all four STAR proteins. Red font indicates a position where only Sam68 differs from the SF1 sequence. A color version of this figure is available at www.landesbioscience.com/curie.

Conservation of RNA Contact Residues between SF1 and STAR Proteins

Structure based homology modeling reveals a high degree of conservation in the RNA contact residues between SF1, GLD-1, QKI and HOW (Fig. 4).[63,64] The GLD-1, QKI and HOW residues predicted to contact RNA in the structure-based alignments are 100% identical with each other.[36] Of these, eighteen out of twenty-five (72%) are identical to those in the SF1 structure. All amino acids in the QUA2 region are conserved, suggesting that GLD-1, QKI and HOW recognize nucleotides corresponding to A4 and C5 in an identical fashion. Four out of five amino acids that comprise the U6 binding pocket are conserved, with the lone exception being a substitution of arginine for Leu 155. While at first glance the leucine to arginine substitution would seem a nonconservative substitution, the aliphatic side-chain of arginine can form the same hydrophobic stacking interactions as leucine with the π-orbitals of U6, consistent with experimental evidence revealing that a uridine residue is strongly specified at this position in GLD-1, QKI and HOW. Likewise, a single conservative substitution is present in the A9 binding pocket, where an aspartic acid replaces the glutamate (Glu 149) in SF1. The identity of the amino acids that recognize A8 and C9 are somewhat less well conserved, where Val 183 is replaced with a methionine, Leu 164 is replaced with an alanine and Ile 157 is replaced with a leucine. It should be noted, however, that all of these substitutions are conservative in nature, replacing a hydrophobic residue with another and Ile 177, which forms a specific hydrogen bond with A10, is conserved in GLD-1, QKI and HOW. Biochemical data reveals that GLD-1, QKI and HOW specify an adenosine residue at a similar position within their consensus binding sites (Table 1).[36-39]

Because of the high degree of sequence identity in the RNA contact residues and because the limited substitutions between species do not affect binding specificity experimentally, it is not surprising that all four proteins recognize similar sequence determinants.[36-39,62] The minor variances in binding specificity between GLD-1, QKI and HOW are not explained by this modeling exercise, as the RNA contact amino acids are 100% identical between all three proteins. Future structural and biochemical analyses will be needed to dissect the structural basis for the differences in RNA recognition by these proteins.

In contrast, there is significant variance between the SF1 RNA contact residues and their counterparts in Sam68.[36,65] Only 40% are identical (10/25), with nonconservative substitutions in amino acids that make base-specific contacts. Only two of the five amino acids in QUA2 that contact A4 and C5 are conserved. Notably, Arg 255, which defines the shape of the A4 binding pocket, is a glutamate in Sam68 (Glu 279). Glutamate is significantly smaller than arginine and as such cannot form the same contacts. Moreover, the glutamate for arginine substitution causes a charge reversal that could have a significant impact on the architecture of the pocket and electrostatic environment needed to recognize polyanionic RNA.

Three of the five amino acids that recognize U8 are conserved, with two substitutions: Leu 155 is replaced with a lysine and Arg 160 is replaced with a glutamine. Lysine can form similar hydrophobic π-orbital stacking interactions as leucine in SF1 and arginine in GLD-1, QKI and HOW. Thus this substitution is not expected to significantly compromise U6 recognition. Arg 160 forms a protein backbone to RNA backbone interaction, thus the amino acid side chain does not directly contribute to nucleotide sequence discrimination. The contact amino acids in Sam68 remain consistent with specification of a uridine nucleotide.

A7 is recognized by a base specific hydrogen bond between the N6 exocyclic amine and a glutamate side chain (Glu 149). In Sam68, the equivalent residue is a lysine (Lys 169). This amino acid cannot form the same contact, thus, adenosine specification cannot occur through a similar mechanism. Only one of the four amino acids that form base specific interactions with A8 and C9 are conserved in Sam68. The substitutions are relatively conservative changes from one small hydrophobic amino acid to another. It is not clear how these changes affect specificity.

Because the QUA2 domain is poorly conserved in Sam68 relative to SF1, GLD-1, QKI and HOW, it is possible that this region does not contact RNA at all. SELEX data suggests that Sam68 binds to a four nucleotide specificity determinant (5'-UAAA-3') as opposed to the five or six nucleotide elements recognized by SF1, GLD-1, QKI and

HOW. Also, two basic amino acids within the QUA2 domain (R240 and K241) that have been experimentally demonstrated to contribute to binding affinity and specificity without directly contacting RNA are not conserved in Sam68, but are in GLD-1, QKI and HOW.[66] Together, the evidence suggests that the Sam68 QUA2 domain diverges significantly from the others and as such its role in RNA recognition is not clear.

In contrast, the U6 binding pocket in Sam68 is similar to the other STAR proteins. We suggest that the first U of the Sam68 recognition sequence is equivalent to U6 in the BPS RNA, while the second and third adenosines are equivalent to A7 and A8. Thus, the KH domain alone could drive specific RNA recognition by Sam68. However, because of the amino acid differences described above, specification of A7 must occur through a different mechanism and it is not at all clear how C9 can be replaced by an adenosine and fit within the binding pocket. The structural basis of specific RNA recognition by Sam68 and its variants remains an important question that merits continued investigation.

CONCLUSION

STAR domain proteins bind to short linear RNA sequences with defined sequence specificity. While much biochemical work has been done to address STAR protein binding specificity, little is known about the mechanism by which sequence specificity is achieved. Some insight can be obtained through homology modeling of the KH and QUA2 regions of the STAR domain using the NMR structure of SF1 as a guide. The high level of conservation of the RNA contact residues between SF1, GLD-1, QKI and HOW and the relative similarity of their consensus binding sites, greatly facilitate this effort. Such comparisons fall short of explaining the basis for the remaining differences in affinity and specificity and do not provide a compelling basis for the mechanism of RNA recognition by Sam68 and related proteins, which falls into a different specificity class. There is much left to be understood about RNA recognition by STAR proteins and as such additional structures of STAR-RNA complexes are needed.

A pressing issue is the role of dimerization in defining binding specificity. STAR proteins form stable dimers, mediated by the QUA1 domain and as such contain two subunits capable of associating with RNA independently. Do both subunits bind to the same sequence? If so, regulatory targets might be expected to have two copies of the consensus binding site. Or does each subunit bind to RNA with different specificity? If so, a bipartite recognition element comprised of different sequences, as observed by SELEX for QKI, may in fact represent the required element in vivo. The shape of the dimer and the path of the RNA through the dimer will likely influence the energetics of RNA recognition. Resolution of these issues awaits functional studies and the determination of the high resolution structure of a ternary complex between a STAR domain dimer and its cognate RNA target sequence.

NOTE ADDED IN PROOF

Williamson and colleagues recently published the high resolution crystal structure of the Qua1 dimerization domain from GLD-1.[67] The structure reveals that this domain folds into a helix-turn-helix motif. The two protomers cross at a 90 degree angle to form the dimer interface. Mutations within the Qua1 domain affect activity by destroying the fold of the helix-turn-helix motif, rather than perturbing the interactions in the interface.

ACKNOWLEDEGMENTS

The authors thank Ruth Zearfoss and John Pagano for critical comments on the manuscript. S.P.R. is supported by NIH R01 GM081422, NIH R21 NS059380 and a Basil O'Connor Starter Scholar Award from the March of Dimes.

REFERENCES

1. Farley BM, Ryder SP. Regulation of maternal mRNAs in early development. Crit Rev Biochem Mol Biol 2008; 43:135-162.
2. Moore MJ. From birth to death: the complex lives of eukaryotic mRNAs. Science 2005; 309:1514-1518.
3. Keene JD, Lager PJ. Post-transcriptional operons and regulons co-ordinating gene expression. Chromosome Res 2005; 13:327-337.
4. Verdel A, Jia S, Gerber S et al. RNAi-mediated targeting of heterochromatin by the RITS complex. Science 2004; 303:672-676.
5. Green R, Noller HF. Ribosomes and translation. Annu Rev Biochem 1997; 66:679-716.
6. Spirin AS. "Masked" forms of mRNA. Curr Top Dev Biol 1966; 1:1-38.
7. Hudson BP, Martinez-Yamout MA, Dyson HJ et al. Recognition of the mRNA AU-rich element by the zinc finger domain of TIS11d. Nat Struct Mol Biol 2004; 11:257-264.
8. Liu Z, Luyten I, Bottomley MJ et al. Structural basis for recognition of the intron branch site RNA by splicing factor 1. Science 2001; 294:1098-1102.
9. Wang X, Zamore PD, Hall TM. Crystal structure of a Pumilio homology domain. Mol Cell 2001; 7:855-865.
10. Edwards TA, Pyle SE, Wharton RP et al. Structure of Pumilio reveals similarity between RNA and peptide binding motifs. Cell 2001; 105:281-289.
11. Chao JA, Patskovsky Y, Almo SC et al. Structural basis for the coevolution of a viral RNA-protein complex. Nat Struct Mol Biol 2008; 15:103-105.
12. Vargason JM, Szittya G, Burgyan J et al. Size selective recognition of siRNA by an RNA silencing suppressor. Cell 2003; 115:799-811.
13. Valegard K, Murray JB, Stockley PG et al. Crystal structure of an RNA bacteriophage coat protein-operator complex. Nature 1994; 371:623-626.
14. Oubridge C, Ito N, Evans PR et al. Crystal structure at 1.92 A resolution of the RNA-binding domain of the U1A spliceosomal protein complexed with an RNA hairpin. Nature 1994; 372:432-438.
15. Agalarov SC, Sridhar Prasad G, Funke PM et al. Structure of the S15,S6,S18-rRNA complex: assembly of the 30S ribosome central domain. Science 2000; 288:107-113.
16. Sonoda J, Wharton RP. Recruitment of Nanos to hunchback mRNA by Pumilio. Genes Dev 1999; 13:2704-2712.
17. Berglund JA, Abovich N, Rosbash M. A cooperative interaction between U2AF65 and mBBP/SF1 facilitates branchpoint region recognition. Genes Dev 1998; 12:858-867.
18. Vernet C, Artzt K. STAR, a gene family involved in signal transduction and activation of RNA. Trends Genet 1997; 13:479-484.
19. Vega-Sanchez ME, Zeng L, Chen S et al. SPIN1, a K homology domain protein negatively regulated and ubiquitinated by the E3 ubiquitin ligase SPL11, is involved in flowering time control in rice. Plant Cell 2008; 20:1456-1469.
20. Sidman RL, Dickie MM, Appel SH. Mutant Mice (Quaking and Jimpy) with Deficient Myelination in the Central Nervous System. Science 1964; 144:309-311.
21. Francis R, Barton MK, Kimble J et al. gld-1, a tumor suppressor gene required for oocyte development in Caenorhabditis elegans. Genetics 1995; 139:579-606.
22. Francis R, Maine E, Schedl T. Analysis of the multiple roles of gld-1 in germline development: interactions with the sex determination cascade and the glp-1 signaling pathway. Genetics 1995; 139:607-630.

23. Jones AR, Francis R, Schedl T. GLD-1, a cytoplasmic protein essential for oocyte differentiation, shows stage- and sex-specific expression during Caenorhabditis elegans germline development. Dev Biol 1996; 180:165-183.
24. Taylor SJ, Shalloway D. An RNA-binding protein associated with Src through its SH2 and SH3 domains in mitosis. Nature 1994; 368:867-871.
25. Lock P, Fumagalli S, Polakis P et al. The human p62 cDNA encodes Sam68 and not the RasGAP-associated p62 protein. Cell 1996; 84:23-24.
26. Ebersole TA, Chen Q, Justice MJ et al. The quaking gene product necessary in embryogenesis and myelination combines features of RNA binding and signal transduction proteins. Nat Genet 1996; 12:260-265.
27. Lin Q, Taylor SJ, Shalloway D. Specificity and determinants of Sam68 RNA binding. Implications for the biological function of K homology domains. J Biol Chem 1997; 272:27274-27280.
28. Baehrecke EH. who encodes a KH RNA binding protein that functions in muscle development. Development 1997; 124:1323-1332.
29. Zaffran S, Astier M, Gratecos D et al. The held out wings (how) Drosophila gene encodes a putative RNA-binding protein involved in the control of muscular and cardiac activity. Development 1997; 124:2087-2098.
30. Nabel-Rosen H, Dorevitch N, Reuveny A et al. The balance between two isoforms of the Drosophila RNA-binding protein how controls tendon cell differentiation. Mol Cell 1999; 4:573-584.
31. Aberg K, Saetre P, Jareborg N et al. Human QKI, a potential regulator of mRNA expression of human oligodendrocyte-related genes involved in schizophrenia. Proc Natl Acad Sci USA 2006; 103:7482-7487.
32. Haroutunian V, Katsel P, Dracheva S et al. The human homolog of the QKI gene affected in the severe dysmyelination "quaking" mouse phenotype: downregulated in multiple brain regions in schizophrenia. Am J Psychiatry 2006; 163:1834-1837.
33. Chen T, Richard S. Structure-function analysis of Qk1: a lethal point mutation in mouse quaking prevents homodimerization. Mol Cell Biol 1998; 18:4863-4871.
34. Chen T, Damaj BB, Herrera C et al. Self-association of the single-KH-domain family members Sam68, GRP33, GLD-1 and Qk1: role of the KH domain. Mol Cell Biol 1997; 17:5707-5718.
35. Ryder SP, Frater L, Abramovitz DL et al. RNA target specificity of the STAR/GSG domain post-transcriptional regulatory protein GLD-1. Nat Struct Mol Biol 2004; 11:20-28.
36. Ryder SP, Frater LA, Abramovitz DL et al. RNA target specificity of the STAR/GSG domain post-transcriptional regulatory protein GLD-1. Nat Struct Mol Biol 2004; 11:20-28.
37. Israeli D, Nir R, Volk T. Dissection of the target specificity of the RNA-binding protein HOW reveals dpp mRNA as a novel HOW target. Development 2007; 134:2107-2114.
38. Galarneau A, Richard S. Target RNA motif and target mRNAs of the Quaking STAR protein. Nat Struct Mol Biol 2005; 12:691-698.
39. Ryder SP, Williamson JR. Specificity of the STAR/GSG domain protein Qk1: implications for the regulation of myelination. RNA 2004; 10:1449-1458.
40. Nabel-Rosen H, Volohonsky G, Reuveny A et al. Two isoforms of the Drosophila RNA binding protein, how, act in opposing directions to regulate tendon cell differentiation. Dev Cell 2002; 2:183-193.
41. Larocque D, Pilotte J, Chen T et al. Nuclear retention of MBP mRNAs in the quaking viable mice. Neuron 2002; 36:815-829.
42. Jan E, Motzny CK, Graves LE et al. The STAR protein, GLD-1, is a translational regulator of sexual identity in Caenorhabditis elegans. EMBO J 1999; 18:258-269.
43. Justice MJ, Bode VC. Three ENU-induced alleles of the murine quaking locus are recessive embryonic lethal mutations. Genet Res 1988; 51:95-102.
44. Ryder SP, Williamson JR. Specificity of the STAR/GSG domain protein Qk1: Implications for the regulation of myelination. RNA 2004; 10:1449-1458.
45. Galarneau A, Richard S. The STAR RNA binding proteins GLD-1, QKI, SAM68 and SLM-2 bind bipartite RNA motifs. BMC Mol Biol 2009; 10:47.
46. Doniach T. Activity of the sex-determining gene tra-2 is modulated to allow spermatogenesis in the C. elegans hermaphrodite. Genetics 1986; 114:53-76.
47. Goodwin EB, Okkema PG, Evans TC et al. Translational regulation of tra-2 by its 3′ untranslated region controls sexual identity in C. elegans. Cell 1993; 75:329-339.
48. Kuwabara PE, Okkema PG, Kimble J. tra-2 encodes a membrane protein and may mediate cell communication in the Caenorhabditis elegans sex determination pathway. Mol Biol Cell 1992; 3:461-473.
49. Ebersole T, Rho O, Artzt K. The proximal end of mouse chromosome 17: new molecular markers identify a deletion associated with quakingviable. Genetics 1992; 131:183-190.
50. Kondo T, Furuta T, Mitsunaga K et al. Genomic organization and expression analysis of the mouse qkI locus. Mamm Genome 1999; 10:662-669.

51. Wu J, Zhou L, Tonissen K et al. The quaking I-5 protein (QKI-5) has a novel nuclear localization signal and shuttles between the nucleus and the cytoplasm. J Biol Chem 1999; 274:29202-29210.
52. Li Z, Zhang Y, Li D et al. Destabilization and mislocalization of myelin basic protein mRNAs in quaking dysmyelination lacking the QKI RNA-binding proteins. J Neurosci 2000; 20:4944-4953.
53. Wu JI, Reed RB, Grabowski PJ et al. Function of quaking in myelination: regulation of alternative splicing. Proc Natl Acad Sci USA 2002; 99:4233-4238.
54. Larocque D, Galarneau A, Liu HN et al. Protection of p27(Kip1) mRNA by quaking RNA binding proteins promotes oligodendrocyte differentiation. Nat Neurosci 2005; 8:27-33.
55. Zhao L, Ku L, Chen Y et al. QKI binds MAP1B mRNA and enhances MAP1B expression during oligodendrocyte development. Mol Biol Cell 2006; 17:4179-4186.
56. Saccomanno L, Loushin C, Jan E et al. The STAR protein QKI-6 is a translational repressor. Proc Natl Acad Sci USA 1999; 96:12605-12610.
57. Ainger K, Avossa D, Diana AS et al. Transport and localization elements in myelin basic protein mRNA. J Cell Biol 1997; 138:1077-1087.
58. Lukong KE, Richard S. Sam68, the KH domain-containing superSTAR. Biochim Biophys Acta 2003; 1653:73-86.
59. Maguire ML, Guler-Gane G, Nietlispach D et al. Solution structure and backbone dynamics of the KH-QUA2 region of the Xenopus STAR/GSG quaking protein. J Mol Biol 2005; 348:265-279.
60. Kramer A, Utans U. Three protein factors (SF1, SF3 and U2AF) function in presplicing complex formation in addition to snRNPs. EMBO J 1991; 10:1503-1509.
61. Arning S, Gruter P, Bilbe G et al. Mammalian splicing factor SF1 is encoded by variant cDNAs and binds to RNA. RNA 1996; 2:794-810.
62. Berglund JA, Chua K, Abovich N et al. The splicing factor BBP interacts specifically with the pre-mRNA branchpoint sequence UACUAAC. Cell 1997; 89:781-787.
63. Gasteiger E, Gattiker A, Hoogland C et al. ExPASy: The proteomics server for in-depth protein knowledge and analysis. Nucleic Acids Res 2003; 31:3784-3788.
64. Schwede T, Kopp J, Guex N et al. SWISS-MODEL: An automated protein homology-modeling server. Nucleic Acids Res 2003; 31:3381-3385.
65. Lehmann-Blount KA, Williamson JR. Shape-specific nucleotide binding of single-stranded RNA by the GLD-1 STAR domain. J Mol Biol 2005; 346:91-104.
66. Garrey SM, Cass DM, Wandler AM et al. Transposition of two amino acids changes a promiscuous RNA binding protein into a sequence-specific RNA binding protein. RNA 2008; 14:78-88.
67. Beuck C, Szymczyna BR, Kerkow DE et al. Structure of the GLD-1 homodimerization domain: insights into STAR protein-mediated translational regulation. Structure 2010; 18:377-89

CHAPTER 4

POST-TRANSLATIONAL REGULATION OF STAR PROTEINS AND EFFECTS ON THEIR BIOLOGICAL FUNCTIONS

Claudio Sette*

Abstract   STAR (Signal Transduction and Activation of RNA) proteins owed their name to the presence in their structure of a RNA-binding domain and several hallmarks of their involvement in signal transduction pathways. In many members of the family, the STAR RNA-binding domain (also named GSG, an acronym for GRP33/Sam68/GLD-1) is flanked by regulatory regions containing proline-rich sequences, which serve as docking sites for proteins containing SH3 and WW domains and also a tyrosine-rich region at the C-terminus, which can mediate protein-protein interactions with partners through SH2 domains. These regulatory regions contain consensus sequences for additional modifications, including serine/threonine phosphorylation, methylation, acetylation and sumoylation. Since their initial description, evidence has been gathered in different cell types and model organisms that STAR proteins can indeed integrate signals from external and internal cues with changes in transcription and processing of target RNAs. The most striking example of the high versatility of STAR proteins is provided by Sam68 (KHDRBS1), whose function, subcellular localization and affinity for RNA are strongly modulated by several signaling pathways through specific modifications. Moreover, the recent development of genetic knockout models has unveiled the physiological function of some STAR proteins, pointing to a crucial role of their post-translational modifications in the biological processes regulated by these RNA-binding proteins. This chapter offers an overview of the most updated literature on the regulation of STAR proteins by post-translational modifications and illustrates examples of how signal transduction pathways can modulate their activity and affect biological processes.

*Claudio Sette—Department of Public Health and Cell Biology, University of Rome "Tor Vergata", Via Montpellier, 1, 00133, Rome, Italy. Email: claudio.sette@uniroma2.it

Post-Transcriptional Regulation by STAR Proteins: Control of RNA Metabolism in Development and Disease, edited by Talila Volk and Karen Artzt.
©2010 Landes Bioscience and Springer Science+Business Media.

INTRODUCTION

The STAR family comprises a class of RNA-binding proteins (RBPs) that are evolutionarily conserved from yeast to humans.[1] The two common features of STAR proteins are the presence of a STAR RNA-binding domain (see below) and of several motifs that confer the ability to form protein-protein interactions and to be modified post-translationally. Several STAR proteins, including Sam68, GLD-1, QKI and GRP33, are capable of homodimerizing in the cell[2] and this feature is required for RNA binding and for many of their functions.[2,3] The ability to homodimerize relies on sequences in the STAR domain,[2] indicating that this region mediates protein-protein interactions in addition to RNA binding. On the other hand, motifs disseminated along the whole structure of different STAR proteins allow heteromeric complexes with numerous proteins involved in signal transduction events and RNA processing (Fig. 1A).[3-5] The only exception to this latter feature of STAR proteins is GLD-1, which lacks obvious hallmarks of motifs involved in signaling.[1] Remarkably, the protein-protein interactions engaged in by some STAR family members have been shown to play a role in propagation of signaling events,[3,4] but they also modulate RNA metabolism, as indicated by the reduced affinity of Sam68 for RNA when bound to SH3 domains.[6] An additional layer of plasticity is provided by the many post-translational modifications of STAR proteins in response to activation of various signal transduction pathways. For instance, tyrosine phosphorylation in the C-terminal tail of many STAR proteins reduces their affinity for RNA and impairs homodimerization (Fig. 1A).[1-3] In line with the impairment of RNA binding,[7,8] tyrosine phosphorylation suppresses the effect of the STAR proteins Sam68 and SLM-1 on alternative splicing.[5,9] Similarly, serine/threonine phosphorylation of Sam68 improves its affinity for specific RNAs, thereby modulating alternative splicing of a target pre-mRNAs (Fig. 1A,B).[10] Moreover, recruitment of Sam68 onto the polysomes and its function in translation appears to depend on phosphorylation (Fig. 1A,B).[11,12] Additional post-translational modifications, such as methylation, acetylation and sumoylation, have been shown to affect specific functions of at least one STAR protein, even though a link to a specific biological process has been less firmly established (Fig. 1A).

Thus, several observations indicate that STAR proteins are crucial integrators of signaling events with regulation of RNA metabolism. Without doubts, the most characterized member of the family in this sense is the mammalian Sam68 protein. For this reason, this chapter will begin with reviewing the information available on post-translational modifications affecting Sam68 functions. Next, examples of other STAR proteins regulated by signal transduction pathways will also be discussed.

Sam68: A BRIEF OVERVIEW

Sam68 is a prototypic STAR protein, with a STAR domain of ~200 amino acids, including the maxi-KH domain embedded in the conserved N-terminal QUA1 and C-terminal QUA2 regions that confer homodimerization and RNA-binding specificity properties.[3] Up-stream and downstream of the STAR domain, Sam68 contains three proline-rich sequences (P0-P5) on each side. In addition, this RBP has RG motifs and a C-terminal tail enriched in tyrosine residues that flank a bipartite nuclear localization signal (Fig. 2A).[3] These features allow Sam68 to interact with multiple proteins,[13] leading to the hypothesis that it might function as a scaffold to cluster signaling proteins in

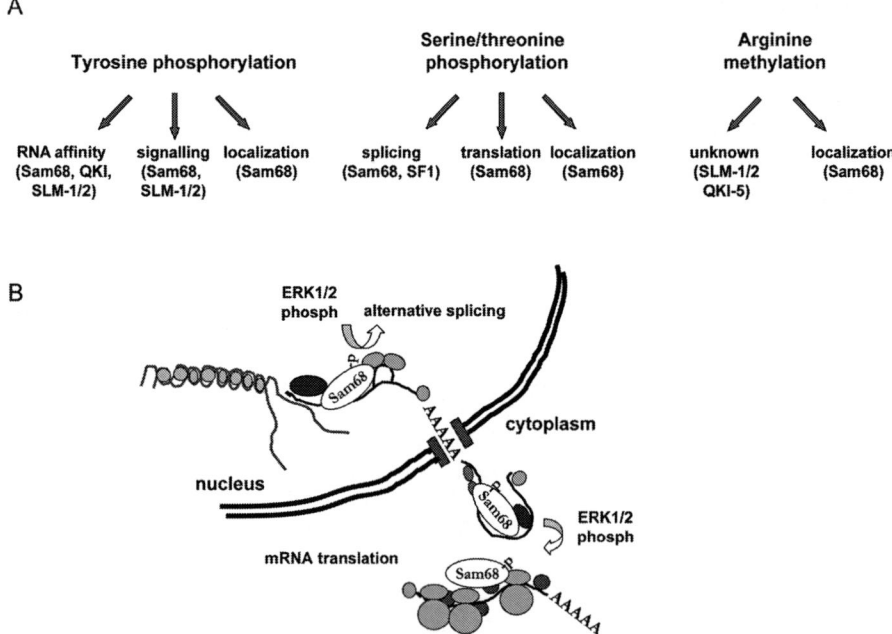

Figure 1. Post-translational modifications of STAR proteins affect their biological functions. A) Schematic representation of the best known post-translational modifications that affect the subcellular localization or activity of specific STAR proteins. B) Schematic representation of the effects of ERK1/2 phosphorylation on the activities of Sam68 in the nucleus and in the cytoplasm. In the nucleus, it was reported that ERK1/2 phosphorylation stimulated the splicing activity of Sam68;[10] in the cytoplasm, it was demonstrated that this phosphorylation event enhanced translation of specific mRNAs in germ cells.[12]

response to specific stimuli.[3,4] More recently, Sam68 has been demonstrated to regulate RNA metabolism at different steps. First, it was shown that it enhanced export and cytoplasmic utilization of viral RNAs,[14,15] complementing the function of the HIV Rev protein. This activity is likely important, because the virus replicates poorly in cells depleted of Sam68 or expressing a dominant-negative Sam68 protein.[16-18] On the other hand, this STAR protein is also implicated in normal nuclear events, such as transcriptional and post-transcriptional regulation of selected cellular transcripts. Through its association with transcription factors or regulators, Sam68 modulates the transcription of target genes.[19-21] This activity could be linked to the effect of Sam68 on alternative splicing,[21,22] since the two process are tightly linked.[23] Indeed, Sam68 was shown to regulate the choice of alternatively spliced exons in *CD44*, *Bcl-x* and a subset of transcripts required for neurogenesis.[5,10,24] Finally, Sam68 was detected in the cytoplasm of primary neurons[25,26] and germ cells,[11,12,27] where it associated with the translation initiation complex eIF4F and the polyribosomes, thereby enhancing translation of a subset of mRNAs (Fig. 1B). This function of Sam68 is likely essential, at least in germ cells, because its genetic ablation leads to defects in germ cell differentiation and to male infertility,[12] see Chapter 5. Thus, the many tasks carried out by Sam68 in different cell types suggest that the activity and subcellular localization of this multifunctional RBP needs to be finely tuned according to the specific requirements of the cell. In line with this hypothesis, Sam68 has been shown to undergo many post-translational modifications under various conditions and in most cases these modifications have an impact on Sam68 function and/or localization. Hence, it is appropriate to begin with a review of the literature illustrating the high versatility of the most studied of the STAR proteins.

REGULATION OF Sam68 FUNCTIONS BY TYROSINE PHOSPHORYLATION

Sam68 was originally identified by virtue of its tyrosine phosphorylation in cells transformed with the Src oncogene[28] and erroneously named p62GAP associated protein.[29] A few years later, two groups independently identified this RBP as a phosphoprotein that associated with Src in mitotic cells (Src Associated in Mitosis protein, of 68 kDa) and renamed it Sam68.[30,31] Sam68 was highly phosphorylated in tyrosine residues in mitotic cells transformed with an oncogenic form of Src and its association with the SH3 and SH2 domains of the kinase was required for phosphorylation. Moreover, these studies showed that Sam68 could bind polyribonucleotides in vitro and suggested that Src might regulate the processing of cellular RNAs through its interaction with this RBP.[30] It was subsequently demonstrated that Sam68 could also associate with the Src-related kinase Fyn in a similar manner.[13] Tyrosine phosphorylation of Sam68 likely promotes the formation of multimolecular complexes that enhance propagation of intracellular signals, as indicated by the active role it plays in antigen-stimulated T-cells.[32-34] Moreover, tyrosine phosphorylation of Sam68 is part of the signaling events triggered by engagement of the insulin and prolactin receptors.[35,36] This suggests that this post-translational modification insures correct hormonal response of target cells. Beside Src family kinases (SFKs), a number of signaling proteins, like PLCγ1, PI3K and the adaptor molecules NCK and GRB2, bind to tyrosine-phosphorylated Sam68.[3,4] In the cell, Sam68 forms two multimolecular complexes of different size and tyrosine phosphorylation stimulates its association with the smaller complex, which also depends on RNA binding and might correlate with the splicing activity of this STAR protein (Huot ME, Vogel G and Richard S; personal communication).

Regulation of signal transduction pathways by Sam68 might play a role in tumorigenesis. Indeed, it was observed that decreased expression of Sam68 delayed the onset of mammary tumors in vivo and that this effect was correlated with increased Src activity in tissue.[37] Remarkably, phosphorylation of Sam68 was increased in specimens from patients affected by breast[38] and prostate cancer.[39] In this latter tumor type, phosphorylation of Sam68 correlated with expression of a truncated form of the c-kit receptor and activation of Src.[39] Notably, Sam68 was previously shown to promote a complex between these kinases.[40] Furthermore, tyrosine phosphorylation of Sam68 was strongly induced by RET/PTC2, an oncogene implicated in thyroid cancers,[41] which acts upstream of SFKs. Finally, Sam68 could play an additional role as modulator of SFK activity in cancer cells. For instance, it was proposed that Sam68 regulates the dynamic assembly of the actin cytoskeleton at the plasma membrane through modulation of the Src signaling pathway and that depletion of Sam68 causes aberrant activation of the Rho small GTPase.[42]

In addition to SFKs, other tyrosine kinases have been shown to interact with and phosphorylate Sam68. The Tec family kinases ITK[43] and BTK[44], which are specifically expressed in T-cells and B-cells, respectively, ZAP-70[45] and BRK[46] all associate with and phosphorylate Sam68 in tyrosine residues. BRK is particularly interesting for its

A

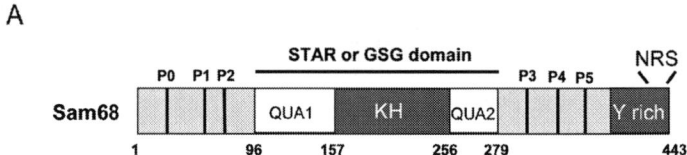

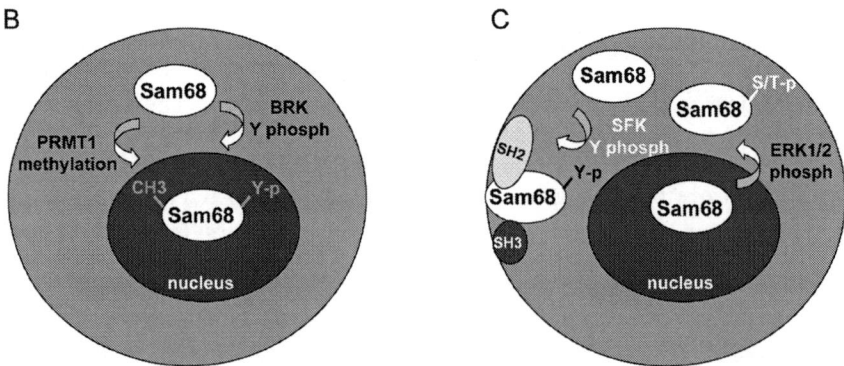

Figure 2. Regulation of Sam68 subcellular localization by post-translational modifications. A) Scheme of the structure of Sam68. P0-P5 identify the position of the proline-rich sequences that are known to interact with SH3 and WW domains; the position of the KH domain for RNA binding and the QUA1 and QUA2 regions of homology with other GSG domains are indicated; Y-rich indicates the region enriched in tyrosine residues that are sites of phosphorylation by BRK, Tec kinases and SFKs; NRS indicates the position of the nuclear retention signal. B) Schematic representation of the stimulation of Sam68 nuclear localization exerted by PRMT1-driven arginine methylation or by BRK-driven tyrosine phosphorylation. C) Schematic representation of the stimulation of Sam68 nuclear export exerted by ERK1/2-driven phosphorylation and of the interaction with signaling proteins containing SH2 or SH3 domains near the plasma membrane exerted by SFK-driven tyrosine phosphorylation.

subcellular localization. Indeed, this tyrosine kinase is mainly localized in the nucleus and accumulates in the same nuclear bodies as Sam68. This suggests that it could be responsible for regulation of the bulk of Sam68, which also resides in the nucleus of interphase cells. Tyrosine phosphorylation by either SFKs or BRK leads to decreased affinity of Sam68 for RNA.[7,46] However, although overexpression studies have shown that the cytoplasmic Fyn can phosphorylate Sam68 and modulate its nuclear activities, such as alternative splicing of target pre-mRNAs[5] or association with the splicing factor YT521-B,[47] it is likely that the endogenous proteins remain separated in the cell by the nuclear envelope. On the other hand, endogenous BRK colocalizes with Sam68 and might represent a better regulator of its nuclear activities. In line with this hypothesis, BRK can repress the ability of Sam68 to export viral RNAs and to promote their cytoplasmic utilization.[15] Moreover, it has been shown that BRK-mediated tyrosine phosphorylation of Sam68 promotes its nuclear translocation in breast cancer cells under stimulation with epidermal growth factor (EGF) (Fig. 2B).[38] Interestingly, BRK is aberrantly regulated in prostate cancer,[48] a tumor type in which Sam68 is up-regulated[49] and hyperphosphorylated[39] and its expression supports growth and survival of the neoplastic cells.[49] These results suggest that aberrant regulation of the BRK/Sam68 pathway might play a role in oncogenesis.

Thus, the evidence above strongly indicates that tyrosine phosphorylation of Sam68 by SFKs or Tec kinases in the cytoplasm (Fig. 2C) and by BRK in the nucleus (Fig. 2B) affects the ability of this RBP to function as a signaling protein and as an RNA modulator.

REGULATION OF Sam68 FUNCTIONS BY SERINE/THREONINE PHOSPHORYLATION

Although tyrosine phosphorylation of Sam68 has been intensively studied in various experimental settings, in several instances it was observed as a consequence of overexpression of SFKs or other tyrosine kinases, or even expression of their constitutively active forms. An attempt to investigate the changes in phosphorylation of the endogenous Sam68 during the cell cycle was originally done by David Shalloway and collaborators.[50] In this study, it was shown that Sam68 is phosphorylated on serine during interphase and in mitotic cells and on threonine only in mitotic cells. The kinase responsible for threonine phosphorylation in mitosis was identified as Cdc2, whereas the kinase(s) responsible for serine phosphorylation in interphase and mitotic cells was not identified. Under these conditions, no tyrosine phosphorylation of Sam68 was detected, possibly due to technical problems.[50] It is conceivable that serine phosphorylation in interphase is exerted by the extracellular regulated kinases 1 and 2 (ERK1/2), members of the mitogen activated protein kinase (MAPK) family. Indeed, it was shown that Sam68 is phosphorylated by ERK1/2 during stimulation of T-lymphoma cells with phorbol ester.[10] Notably, phosphorylation of Sam68 by ERK1/2 affected the splicing activity of this STAR protein (Fig. 1A,B), enhancing the inclusion of the variable exon 5 (v5) in the CD44 mRNA.[10] Since inclusion of this variable exon positively correlates with neoplastic transformation,[51] modulation of the splicing activity of Sam68 by the MAPK-dependent phosphorylation might represent another cancer-related event involving this multifunctional STAR protein. Mechanistically, it was proposed that Sam68 formed a complex with the splicing factor U2AF65, which recognizes the 3′ splice site, thereby enhancing its recruitment on the v5 exon.[52] Phosphorylation of Sam68 by ERK1/2 would favour the dynamic recruitment of other spliceosomal components through changes in affinity of the Sam68/U2AF65 complex for the v5 RNA.[52]

Serine/threonine phosphorylation of Sam68 also occurs in male germ cells undergoing the meiotic divisions.[11] Similarly to somatic cells, phosphorylation was due to the activity of Cdc2 and ERK1/2 and correlated with the localization of Sam68 in the cytoplasm and its association with the polyribosomes (Figs. 1B and 2C). Using specific inhibitors, it was shown that the ERK1/2-mediated phosphorylation was the main regulator of the association of Sam68 with the translational machinery.[11] A subsequent study indicated that translocation of Sam68 to the cytoplasm and its association with polyribosomes was required for translational activation of a subset of mRNAs that are target of this STAR protein in germ cells (Fig. 1B).[12] This effect could be recapitulated in somatic cells by transfecting a constitutively active form of RAS, which led to activation of ERK1/2, enhanced phosphorylation of Sam68 and its association with polyribosomes as in germ cells.[12] Remarkably, since germ cells ablated of Sam68 express lower levels of the proteins encoded by these target mRNAs and Sam68 knockout male mice were sterile and produced few spermatozoa,[12] it is likely that this post-translational regulation of Sam68 is essential for male fertility in vivo.

These observations suggest that serine/threonine phosphorylation of Sam68 mainly affects its RNA-binding activity and the functions related to it.

REGULATION OF Sam68 FUNCTIONS BY METHYLATION

The proline-rich sequences in Sam68 are flanked by RG (arginine-glycine) repeats that are consensus for methylation by the Type I of protein arginine methyltransferases (Type I PRMTs).[53] It was shown that methylation of these RG repeats occurs in vitro by incubation of Sam68 with PRMT1. Proline-rich sequences bind SH3 and WW domains. Interestingly it was demonstrated that methylation of Sam68 in the RG repeats that flanked the proline-rich sequences decreased binding to SH3 domains but not WW domains,[53] suggesting that methylation could affect the choice of partners by Sam68 in the cell. In support of a role for methylation in vivo, Sam68 associated with PRMT1 and was constitutively methylated in live cells.[54] Moreover, methylation was required for efficient nuclear localization of the protein (Figs. 1A and 2B) and for its ability to favour the export of viral RNAs,[54] suggesting that it is an essential modification required for Sam68 function. Decreased methylation of Sam68 might occur under pathological conditions. It was shown that peroxynitrite, a pro-atherogenic substance known to induce endothelial dysfunction, inhibited arginine methylation of this STAR protein. The authors reported that reduced methylation of Sam68 did not affect its RNA binding activity, or its levels of tyrosine phosphorylation. However, they showed that it correlated with increased rate of apoptosis and premature senescence of endothelial cells.[55] By contrast, reduction of RNA-binding by methylation of the Sam68 RG repeats was shown by another group.[56] They suggested that the RG repeats provide an additional RNA-binding motif to STAR proteins, outside of the GSG domain and that methylation can modify the affinity of the RG repeats for RNA.[56] However, since methylation of Sam68 appears to be a constitutive event,[54] it seems unlikely that it impedes RNA binding. Although both studies employed poly-uridine synthetic oligonucleotides to test Sam68 affinity for RNA, they reached opposite conclusions. Thus, more physiological RNA substrates need to be tested to fully understand the influence of methylation on Sam68 RNA-binding affinity.

Remarkably, the ability of Sam68 to interact with PRMT1 is exploited by an oncogene to elicit neoplastic transformation. The MLL-EEN translocation causes Mixed Lineage Leukemia (MLL) in humans. A recent study showed that the SH3 domain of EEN is the only part of this protein required for MLL oncogenesis.[57] A screen for proteins interacting with this SH3 domain identified Sam68 and subsequent experiments demonstrated that recruitment of Sam68 to MLL-EEN by the SH3 domain was crucial to elicit neoplastic transformation.[57] Interestingly, the activity of Sam68 important for cell transformation was its ability to associate with PRMT1. Indeed, direct fusion of PRMT1 with MLL bypassed the requirement of both EEN and Sam68. Thus, MLL-EEN induced transformation through the recruitment of a complex formed by Sam68 and PRMT1 with the SH3 domain of EEN. This complex allows PRMT1 to be recruited to MLL-sensitive promoters and to alter the epigenetic status of the responsive genes, hence modifying gene expression.[57] This study links the scaffold function of Sam68 and its connection with methyltransferases with oncogenesis and highlights how a versatile STAR protein can be used by the cell for unexpected functions, such as those set in motion by a mutated oncogene.

REGULATION OF Sam68 FUNCTIONS BY ACETYLATION AND SUMOYLATION

Another post-translational modification that can affect Sam68 function in the cell is acetylation. It was initially observed that Sam68 associated with the histone acetyltransferase CBP on specific promoters.[19] Few years later, it was observed that Sam68 is preferentially acetylated in breast cancer cell lines as compared to normal breast epithelial cells.[58] Moreover, acetylation of Sam68 positively correlated with its ability to bind poly-uridine synthetic RNA in extracts obtained from these cell lines. The authors also showed that CBP could acetylate Sam68 in vitro, mainly on lysines present in the QUA1 region and the first half of the KH domain, hence increasing RNA binding and that similar results are obtained by overexpressing CBP in transfected cells.[58] However, the role played by acetylation in the biological functions of Sam68 still needs to be addressed.

Sam68 was shown to form complexes with transcriptional regulators and modulate transcription of reporter genes.[19,22] One of the transcriptional targets of Sam68 in normal and cancer cells is cyclin D1.[20,49] This activity of Sam68 might also be subject to regulation by a post-translational modification. It was reported that Sam68 is modified by covalent link with the small ubiquitin-like protein SUMO.[59] This reaction was catalyzed by the SUMO E3 ligase known as PIAS1 and occurred on lysine 96 of Sam68. Since mutation of this acceptor site enhanced the pro-apoptotic activity of Sam68, which is dependent on its RNA-binding activity,[5,20] it is possible that sumoylation affects the binding of Sam68 to its cellular mRNA targets. On the other hand, sumoylation increased the repression of the cyclin D1 promoter by Sam68. A SUMO1-Sam68 fusion protein recapitulated these events, causing stronger repression of cyclin D1 expression and lower levels of apoptosis.[59] Thus, it is possible that sumoylation of endogenous Sam68 finely tunes the ability of this protein to regulate cell cycle progression and apoptosis in response to specific signals. However, the mechanism by which sumoylation affects Sam68 activity is still unknown and more work is needed to ascertain the biological importance of this modification in live cells.

POST-TRANSLATIONAL MODIFICATIONS OF SLM-1 AND SLM-2

The STAR proteins that are more related to Sam68 are SLM-1 and SLM-2 (Sam68 Like Mammalian protein 1 and 2).[60] SLM-2 was independently cloned by another group and named T-STAR and étoile in human and mouse, respectively.[61] These proteins share approximately 65-70% sequence identity with Sam68 in their STAR domain and have similar SH2 and SH3 domain binding sites.[60] Moreover, many of the post-translational modifications described for Sam68 apply also to SLM-1 and SLM-2 (Fig. 1A). However, some differences exist. For instance, SLM-2 was not phosphorylated by SFKs and did not interact with the SH3 domains of several signaling proteins tested,[60] suggesting that it lacked the scaffold function of Sam68 and SLM-1. On the other hand, both SLM-1 and SLM-2 were phosphorylated by the nuclear BRK and this modification decreased their affinity for synthetic RNA in vitro.[8] Another feature in common between Sam68 ans SLM-2 was their methylation by PRMT1,[54] which was suggested to decrease their affinity for RNA in the same in vitro study in which Sam68 was analysed.[56] In terms of biological roles, they also share many features with Sam68. SLM proteins synergize with the HIV protein Rev in stimulating gene expression of responsive genes, but, differently from Sam68, they could not substitute for the Rev function.[62] Both SLM-1 and SLM-2 function in alternative splicing like Sam68,[9,63] but only SLM-1 activity is inhibited by SFK-mediated tyrosine phosphorylation.[9] Thus, SLM-1 and 2 appear to be very close homologues of Sam68 that share many of its features and activities. Since Sam68 knockout mice are viable,[64] it is possible that they compensate for lack of their cousin STAR protein and support viability of these mice.

POST-TRANSLATIONAL MODIFICATIONS OF THE QKI PROTEINS

The mouse quaking (*Qk*) gene is essential for central nervous system (CNS) myelination and for survival of the early embryo.[65] The proteins encoded by this gene (QKI5, QKI6 and QKI7) show the typical structure of STAR proteins, combining RNA-binding with signal transduction properties and are expressed in myelin-forming cells and in astrocytes.[66] QKI6 and QKI7 are localized to cytoplasm, whereas a nuclear localization signal in the C-terminus of QKI5 allows its import in the nucleus.[67] Mutations in the *Qk* gene that are compatible with life, named *qk(v)*, cause severe dysmyelination in mice, which correlates with aberrant expression of the QKI proteins in different subsets of glial cells.[66] On the other hand, a mutation in the STAR domain that impairs homodimerization is lethal in vivo,[68] highlighting the importance of this self-interaction in the function of STAR proteins. QKI proteins mediate post-transcriptional regulation of mRNAs encoding several proteins involved in the formation of the myelin sheet. For instance, the nuclear QKI5 isoform regulates alternative splicing of exon 12 in the myelin-associated glycoprotein (MAG) pre-mRNA and this splicing event is altered in the quaking mice.[69] Moreover, the QKI proteins are required for the stabilization and export of the mRNAs encoding the myelin basic protein (MBP) isoforms, thereby causing accumulation of this protein in cells undergoing myelinogenesis.[70,71] Notably, the interaction between QKI and MBP mRNA is regulated by SFKs in the developing CNS. It was shown that, similarly to Sam68, phosphorylation of the C-terminal tyrosine residues in QKI proteins by Src or Fyn decreased their affinity for target RNA in vivo and in vitro (Fig. 1A).[72] When examined in the developing brain, tyrosine phosphorylation of QKI proteins was maximal at day 7 postnatal and rapidly declined from day 7 to day 20, concomitantly with the strong induction in MBP mRNA and protein levels and with myelinogenesis.[72] In addition, expression of Fyn was elevated in oligodendrocyte precursor cells, whereas the activity of this SFK declined later on during myelin accumulation.[73] Notably, Fyn and QKI activity seemed also to antagonistically regulate alternative splicing of MBP mRNA isoforms.[73] These observations strongly indicated that post-translational regulation of QKI proteins affected the accumulation of one component of the myelin sheet in developing neurons.

In addition to phosphorylation, QKI5 was weakly methylated in arginines in vivo (Fig. 1A). However, in contrast to Sam68, this STAR protein did not associate with protein methyltransferase activity and with PRMT1.[54] These observations suggested that QKI-5 is a target for a different methyltransferase in vivo and that this post-translational modification might not require a stable interaction.[54] The consequences of methylation on QKI-5 activity and whether or not other QKI isoforms are methylated in vivo, remain to be established. Moreover, no information on additional post-translational modifications of these STAR proteins has been reported, suggesting that more studies are required to fully understand QKI regulation in vivo.

POST-TRANSLATIONAL MODIFICATIONS OF SF1

The more distantly related STAR protein SF1 (Splicing Factor 1) is a branchpoint binding protein involved in early steps of the splicing reaction.[1,74,75] Differently from Sam68, SLMs and QKI proteins, the proline-rich sequences of SF1 were shown to preferentially interact with the tyrosine kinase Abl rather than with SFKs.[1] Moreover, these regions allow SF1 to interact with WW domains of proteins that may link transcription to pre-mRNA processing.[76] SF1 binds the intron branch site and associates with the splicing regulator U2AF65.[77] Interestingly, the cGMP-dependent protein kinase-I (PKG-I) phosphorylates SF1 at serine 20, thereby impairing its interaction with U2AF65 and spliceosomal assembly in neuronal cells.[78] This observation suggested that signaling events that alter the levels of cGMP in neurons can exert an effect on splicing through regulation of SF1. On the other hand, the protein kinase KIS (kinase interacting stathmin) interacts with SF1 and phosphorylates two serine residues (aa 80 and 82) flanked by a proline (SPSP motif) and these phosphorylation events enhanced the interaction of SF1 with U2AF65.[79] Thus, the splicing activity of SF1 can be tightly regulated through phosphorylation by different kinases that exert opposite effects on the recruitment of U2AF65 by this STAR protein (Fig. 1A).

CONCLUSION

Since their first identification, the structural features of members of the STAR family suggested a role for these proteins at the crossroad between signal transduction pathways and RNA metabolism.[1] Much work has been done in the past decade to confirm this initial hypothesis. Several STAR proteins have been demonstrated to be post-translationally modified in response to external and internal cues (Fig. 1A). Moreover, examples of their ability to bind RNA and to affect transcriptional and post-transcriptional processing of target mRNAs have been illustrated. Nevertheless, much remains to be done, especially for those proteins lacking genetic support for their in vivo function(s). In addition, the biological influence of some modifications, such as sumoylation or acetylation for Sam68, needs to be further investigated. Another aspect to be clarified is the potential redundancy of Sam68, SLM-1 and SLM-2. These STAR proteins are highly homologous and share many features in terms of signal transduction and RNA activities. However, some differences at the level of post-translational modifications have been demonstrated (i.e., association with PRMT activity for SLM-1; tyrosine phosphorylation by SFKs for SLM-2). Notably, although Sam68 has been implicated in many crucial biological processes, cells can proliferate and survive without it and, in addition, knockout mice can develop and live throughout adulthood, albeit at lower rates than their littermates.[64] It is possible that SLM-1 and SLM-2 constitute a backup activity that supports cell viability in the absence of Sam68 and that part of the defects observed in the knockout animals are due to the slightly different post-translational modifications occurring in the SLM proteins. Finally, post-translational modifications of other members have not been reported yet, but hints indicate that they might occur. For instance, the *D. melanogaster* HOW(L) nuclear protein contains the HPYR motif that is conserved in the nuclear localization signal of other nuclear STAR proteins like QKI-5 and Sam68 and its mutation caused mislocalization of HOW(L) in the cytoplasm.[80] This observation suggests that, similarly to Sam68 (Fig. 2B), the nuclear localization of HOW(L) could be regulated by post-translational modifications.

Thus, future studies are warranted to fully determine the impact of post-translational control on the many activities played by the STAR proteins and to elucidate the many functions of this highly regulated family of RBPs in cells and live organisms.

ACKNOWLEDGEMENTS

The author would like to thank Drs Stéphane Richard and Marc-Etienne Huot for communication of results before publication and Drs Talila Volk and Simona Pedrotti for helpful suggestions. The work in the laboratory of Prof. Claudio Sette was supported by Grants from the "Associazione Italiana Ricerca sul Cancro" (AIRC), the "Association for International Cancer Research" (AICR), the Istituto Superiore della Sanità (ISS Project n.527/B/3A/5) and Telethon.

REFERENCES

1. Vernet C, Artzt K. STAR, a gene family involved in signal transduction and activation of RNA. Trends Genet 1997; 13:479-84.
2. Chen T, Damaj BB, Herrera C et al. Self-association of the single-KH-domain family members Sam68, GRP33, GLD-1 and Qk1: role of the KH domain. Mol Cell Biol 1997; 17:5707-18.
3. Lukong KE, Richard S. Sam68, the KH domain-containing superSTAR. Bioch Biophys Acta 2003; 1653:73-86.
4. Najib S, Martín-Romero C, González-Yanes C et al. Role of Sam68 as an adaptor protein in signal transduction. Cell Mol Life Sci 2005; 62:36-43.
5. Paronetto MP, Achsel T, Massiello A et al. The RNA-binding protein Sam68 modulates the alternative splicing of Bcl-x. J Cell Biol 2007; 176:929-939.
6. Taylor SJ, Anafi M, Pawson T et al. Functional interaction between c-Src and its mitotic target, Sam 68. J Biol Chem 1995; 270:10120-4.
7. Wang LL, Richard S, Shaw AS. P62 association with RNA is regulated by tyrosine phosphorylation. J Biol Chem 1995; 270:2010-2013.
8. Haegebarth A, Heap D, Bie W et al. The nuclear tyrosine kinase BRK/Sik phosphorylates and inhibits the RNA-binding activities of the Sam68-like mammalian proteins SLM-1 and SLM-2. J Biol Chem 2004; 279:54398-54404.
9. Stoss O, Novoyatleva T, Gencheva M et al. p59(fyn)-mediated phosphorylation regulates the activity of the tissue-specific splicing factor rSLM-1. Mol Cell Neurosci 2004; 27:8-21.
10. Matter N, Herrlich P, Konig H. Signal-dependent regulation of splicing via phosphorylation of Sam68. Nature 2002; 420:691-695.
11. Paronetto MP, Zalfa F, Botti F et al. The nuclear RNA-binding protein Sam68 translocates to the cytoplasm and associates with the polysomes in mouse spermatocytes. Mol Biol Cell 2006; 17:14-24.
12. Paronetto MP, Messina V, Bianchi E et al. Sam68 regulates translation of target mRNAs in male germ cells, necessary for mouse spermatogenesis. J Cell Biol 2009; 185:235-249.
13. Richard S, Yu D, Blumer KJ et al. Association of p62, a multifunctional SH2- and SH3-domain-binding protein, with src family tyrosine kinases, Grb2 and phospholipase C gamma-1. Mol Cell Biol 1995; 15:186-197.
14. Li J, Liu Y, Kim BO et al. Direct participation of Sam68, the 68-kilodalton Src-associated protein in mitosis, in the CRM1-mediated Rev nuclear export pathway. J Virol 2002; 76:8374-82.
15. Coyle JH, Guzik BW, Bor YC et al. Sam68 enhances the cytoplasmic utilization of intron-containing RNA and is functionally regulated by the nuclear kinase Sik/BRK. Mol Cell Biol 2003; 23:92-103.
16. Modem S, Badri KR, Holland TC et al. Sam68 is absolutely required for Rev function and HIV-1 production. Nucleic Acids Res 2005;33:873-9.
17. Cochrane A. Inhibition of HIV-1 gene expression by Sam68 Delta C: multiple targets but a common mechanism? Retrovirology 2009; 6:22.
18. Henao-Mejia J, Liu Y, Park IW et al. Suppression of HIV-1 Nef translation by Sam68 mutant-induced stress granules and nef mRNA sequestration. Mol Cell 2009; 33:87-96.
19. Hong W, Resnick RJ, Rakowski C et al. Physical and functional interaction between the transcriptional cofactor CBP and the KH domain protein Sam68. Mol Cancer Res 2002; 1:48-55.

20. Taylor SJ, Resnick RJ, Shalloway D. Sam68 exerts separable effects on cell cycle progression and apoptosis. BMC Cell Biol 2004; 5:5.
21. Batsche E, Ianiv M, Muchardt C. The human SWI/SNF subunit Brm is a regulator of alternative splicing. Nat Struct Mol Biol 2006; 13:22-29.
22. Rajan P, Gaughan L, Dalgliesh C et al. The RNA-binding and adaptor protein Sam68 modulates signal-dependent splicing and transcriptional activity of the androgen receptor. J Pathol 2008; 215:67-77.
23. Kornblihtt AR. Coupling transcription and alternative splicing. Adv Exp Med Biol 2007; 623:175-189.
24. Chawla G, Lin CH, Han A et al. Sam68 regulates a set of alternatively spliced exons during neurogenesis. Mol Cell Biol 2009; 29:201-13.
25. Ben Fredj N, Grange J, Sadoul R et al. Depolarization-induced translocation of the RNA-binding protein Sam68 to the dendrites of hippocampal neurons. J Cell Sci 2004; 117:1079-1090.
26. Grange J, Belly A, Dupas S et al. Specific interaction between Sam68 and neuronal mRNAs: implication for the activity-dependent biosynthesis of elongation factor eEF1A. J Neurosci Res 2009; 87:12-25.
27. Paronetto MP, Bianchi E, Geremia R et al. Dynamic expression of the RNA-binding protein Sam68 during mouse pre-implantation development. Gene Expr Patterns 2008; 8:311-322.
28. Wong G, Müller O, Clark R et al. Molecular cloning and nucleic acid binding properties of the GAP-associated tyrosine phosphoprotein p62. Cell 1992; 69:551-8.
29. Lock P, Fumagalli S, Polakis P et al. The human p62 cDNA encodes Sam68 and not the RasGAP-associated p62 protein. Cell 1996; 84:23-24.
30. Taylor SJ, Shalloway D. An RNA-binding protein associated with Src through its SH2 and SH3 domains in mitosis. Nature 1994; 368:867-871.
31. Fumagalli S, Totty NF, Hsuan JJ et al. A target for Src in mitosis. Nature 1994; 368:871-874.
32. Jabado N, Pallier A, Le Deist F et al. CD4 ligands inhibit the formation of multifunctional transduction complexes involved in T-cell activation. J Immunol 1997; 158:94-103.
33. Fusaki N, Iwamatsu A, Iwashima M et al. Interaction between Sam68 and Src family tyrosine kinases, Fyn and Lck, in T-cell receptor signaling. J Biol Chem 1997; 272:6214-6219.
34. Trüb T, Frantz JD, Miyazaki M et al. The role of a lymphoid-restricted, Grb2-like SH3-SH2-SH3 protein in T-cell receptor signaling. J Biol Chem 1997; 272:894-902.
35. Sánchez-Margalet V, Najib S. Sam68 is a docking protein linking GAP and PI3K in insulin receptor signaling. Mol Cell Endocrinol 2001; 183:113-121.
36. Fresno Vara JA, Cáceres MA et al. Src family kinases are required for prolactin induction of cell proliferation. Mol Biol Cell 2001; 12:2171-2183.
37. Richard S, Vogel G, Huot ME et al. Sam68 haploinsufficiency delays onset of mammary tumorigenesis and metastasis. Oncogene 2008; 27:548-556.
38. Lukong KE, Larocque D, Tyner AL et al. Tyrosine phosphorylation of sam68 by breast tumor kinase regulates intranuclear localization and cell cycle progression. J Biol Chem 2005; 280:38639-38647.
39. Paronetto MP, Farini D, Sammarco I et al. Expression of a truncated form of the c-Kit tyrosine kinase receptor and activation of Src kinase in human prostatic cancer. Am J Pathol 2004; 164:1243-1251.
40. Paronetto MP, Venables JP, Elliott DJ et al. Tr-kit promotes the formation of a multimolecular complex composed by Fyn, PLCgamma1 and Sam68. Oncogene 2003; 22:8707-8715.
41. Gorla L, Cantù M, Miccichè F et al. RET oncoproteins induce tyrosine phosphorylation changes of proteins involved in RNA metabolism. Cell Signal 2006; 18:2272-2282.
42. Huot ME, Brown CM, Lamarche-Vane N et al. An adaptor role for cytoplasmic Sam68 in modulating Src activity during cell polarization. Mol Cell Biol 2009; 29:1933-1943.
43. Andreotti AH, Bunnell SC, Feng S et al. Regulatory intramolecular association in a tyrosine kinase of the Tec family. Nature 1997; 385:93-97.
44. Guinamard R, Fougereau M, Seckinger P. The SH3 domain of Bruton's tyrosine kinase interacts with Vav, Sam68 and EWS. Scand J Immunol 1997; 45:587-595.
45. Lang V, Mège D, Semichon M et al. A dual participation of ZAP-70 and scr protein tyrosine kinases is required for TCR-induced tyrosine phosphorylation of Sam68 in Jurkat T-cells. Eur J Immunol 1997; 27:3360-3367.
46. Derry JJ, Richard S, Valderrama Carvajal H et al. Sik (BRK) phosphorylates Sam68 in the nucleus and negatively regulates its RNA binding ability. Mol Cell Biol 2000; 20:6114-6126.
47. Hartmann AM, Nayler O, Schwaiger FW et al. The interaction and colocalization of Sam68 with the splicing-associated factor YT521-B in nuclear dots is regulated by the Src family kinase p59(fyn). Mol Biol Cell 1999; 10:3909-3926.
48. Derry JJ, Prins GS, Ray V et al. Altered localization and activity of the intracellular tyrosine kinase BRK/Sik in prostate tumor cells. Oncogene 2003; 22:4212-4220.
49. Busà R, Paronetto MP, Farini D et al. The RNA-binding protein Sam68 contributes to proliferation and survival of human prostate cancer cells. Oncogene 2007; 26:4372-4382.
50. Resnick RJ, Taylor SJ, Lin Q et al. Phosphorylation of the Src substrate Sam68 by Cdc2 during mitosis. Oncogene 1997; 15:1247-1253.

51. Lee SC, Harn HJ, Lin TS et al. Prognostic significance of CD44v5 expression in human thymic epithelial neoplasms. Ann Thorac Surg 2003; 76:213-218.
52. Tisserant A, König H. Signal-regulated Pre-mRNA occupancy by the general splicing factor U2AF. PLoS One 2008; 3:e1418.
53. Bedford MT, Frankel A, Yaffe MB et al. Arginine methylation inhibits the binding of proline-rich ligands to Src homology 3, but not WW, domains. J Biol Chem 2000; 275:16030-16036.
54. Côté J, Boisvert FM, Boulanger MC et al. Sam68 RNA binding protein is an in vivo substrate for protein arginine N-methyltransferase 1. Mol Biol Cell 2003; 14:274-287.
55. Polotskaia A, Wang M, Patschan S et al. Regulation of arginine methylation in endothelial cells: role in premature senescence and apoptosis. Cell Cycle 2007; 6:2524-30.
56. Rho J, Choi S, Jung CR et al. Arginine methylation of Sam68 and SLM proteins negatively regulates their poly(U) RNA binding activity. Arch Biochem Biophys 2007; 466:49-57.
57. Cheung N, Chan LC, Thompson A et al. Protein arginine-methyltransferase-dependent oncogenesis. Nat Cell Biol 2007; 9:1208-1215.
58. Babic I, Jakymiw A, Fujita DJ. The RNA binding protein Sam68 is acetylated in tumor cell lines and its acetylation correlates with enhanced RNA binding activity. Oncogene 2004; 23:3781-3789.
59. Babic I, Cherry E, Fujita DJ. SUMO modification of Sam68 enhances its ability to repress cyclin D1 expression and inhibits its ability to induce apoptosis. Oncogene 2006; 25:4955-4964.
60. Di Fruscio M, Chen T, Richard S. Characterization of Sam68-like mammalian proteins SLM-1 and SLM-2: SLM-1 is a Src substrate during mitosis. Proc Natl Acad Sci USA 1999; 96:2710-2715.
61. Venables JP, Vernet C, Chew SL et al. T-STAR/ETOILE: a novel relative of SAM68 that interacts with an RNA-binding protein implicated in spermatogenesis. Hum Mol Genet 1999; 8:959-969.
62. Reddy TR, Suhasini M, Xu W et al. A role for KH domain proteins (Sam68-like mammalian proteins and quaking proteins) in the post-transcriptional regulation of HIV replication. J Biol Chem 2002; 277:5778-5784.
63. Stoss O, Olbrich M, Hartmann AM et al. The STAR/GSG family protein rSLM-2 regulates the selection of alternative splice sites. J Biol Chem 2001; 276:8665-8673.
64. Richard S, Torabi N, Franco GV et al. Ablation of the Sam68 RNA binding protein protects mice from age-related bone loss. PLoS Genet 2005; 1(6):e74.
65. Ebersole TA, Chen Q, Justice MJ et al. The quaking gene product necessary in embryogenesis and myelination combines features of RNA binding and signal transduction proteins. Nat Genet 1996; 12:260-265.
66. Hardy RJ, Loushin CL, Friedrich VL Jr et al. Neural cell type-specific expression of QKI proteins is altered in quakingviable mutant mice. J Neurosci 1996; 16:7941-7949.
67. Wu J, Zhou L, Tonissen K et al. The quaking I-5 protein (QKI-5) has a novel nuclear localization signal and shuttles between the nucleus and the cytoplasm. J Biol Chem 1999; 274:29202-29210.
68. Chen T, Richard S. Structure-function analysis of Qk1: a lethal point mutation in mouse quaking prevents homodimerization. Mol Cell Biol 1998; 18:4863-4871.
69. Wu JI, Reed RB, Grabowski PJ et al. Function of quaking in myelination: regulation of alternative splicing. Proc Natl Acad Sci USA 2002; 99:4233-4238.
70. Li Z, Zhang Y, Li D et al. Destabilization and mislocalization of myelin basic protein mRNAs in quaking dysmyelination lacking the QKI RNA-binding proteins. J Neurosci 2000; 20:4944-4953.
71. Larocque D, Pilotte J, Chen T et al. Nuclear retention of MBP mRNAs in the quaking viable mice. Neuron 2002; 36:815-829.
72. Zhang Y, Lu Z, Ku L et al. Tyrosine phosphorylation of QKI mediates developmental signals to regulate mRNA metabolism. EMBO J 2003; 22:1801-1810.
73. Lu Z, Ku L, Chen Y et al. Developmental abnormalities of myelin basic protein expression in fyn knock-out brain reveal a role of Fyn in post-transcriptional regulation. J Biol Chem 2005; 280:389-395.
74. Krämer A. Purification of splicing factor SF1, a heat-stable protein that functions in the assembly of a presplicing complex. Mol Cell Biol 1992; 12:4545-4552.
75. Liu Z, Luyten I, Bottomley MJ et al. Structural basis for recognition of the intron branch site RNA by splicing factor 1. Science 2001; 294:1098-1102.
76. Lin KT, Lu RM, Tarn WY. The WW domain-containing proteins interact with the early spliceosome and participate in pre-mRNA splicing in vivo. Mol Cell Biol 2004; 24:9176-9185.
77. Berglund JA, Abovich N, Rosbash M. A cooperative interaction between U2AF65 and mBBP/SF1 facilitates branchpoint region recognition. Genes Dev 1998; 12:858-867.
78. Wang X, Bruderer S, Rafi Z et al. Phosphorylation of splicing factor SF1 on Ser20 by cGMP-dependent protein kinase regulates spliceosome assembly. EMBO J 1999; 18:4549-4559.
79. Manceau V, Swenson M, Le Caer JP et al. Major phosphorylation of SF1 on adjacent Ser-Pro motifs enhances interaction with U2AF65. FEBS J 2006; 273:577-587.
80. Nabel-Rosen H, Dorevitch N, Reuveny A et al. The balance between two isoforms of the Drosophila RNA-binding protein how controls tendon cell differentiation. Mol Cell 1999; 4:573-584.

CHAPTER 5

EXPRESSION AND FUNCTIONS OF THE STAR PROTEINS Sam68 AND T-STAR IN MAMMALIAN SPERMATOGENESIS

Ingrid Ehrmann and David J. Elliott*

Abstract: Spermatogenesis is one of the few major developmental pathways which are still ongoing in the adult. In this chapter we review the properties of Sam68 and T-STAR, which are the STAR proteins functionally implicated in mammalian spermatogenesis. Sam68 is a ubiquitously expressed member of the STAR family, but has an essential role in spermatogenesis. Sam68 null mice are male infertile and at least in part this is due to a failure in important translational controls that operate during and after meiosis. The homologous T-STAR protein has a much more restricted anatomic expression pattern than Sam68, with highest levels seen in the testis and the developing brain. The focus of this chapter is the functional role of Sam68 and T-STAR proteins in male germ cell development. Since these proteins are known to have many cellular functions we extrapolate from other cell types and tissues to speculate on each of their likely functions within male germ cells, including control of alternative pre-mRNA splicing patterns in male germ cells.

GENE EXPRESSION CONTROL IN SPERMATOGENESIS

Male germ cell development (spermatogenesis) takes place in the testis and involves the conversion of diploid stem cells into motile haploid spermatozoa that are capable of swimming to and fertilising an egg. Spermatogenesis is very prolific-10^8 sperm cells are produced in an adult mouse/day and the testis is one of the major sites of development ongoing in the adult. A micrograph of a section of rat testis stained for Sam68 protein

*Corresponding Author: David J. Elliott—Institute of Human Genetics, Newcastle University, International Centre for Life, Central Parkway, Newcastle NE1 3BZ, England.
Email: david.elliott@ncl.ac.uk

Post-Transcriptional Regulation by STAR Proteins: Control of RNA Metabolism in Development and Disease, edited by Talila Volk and Karen Artzt.

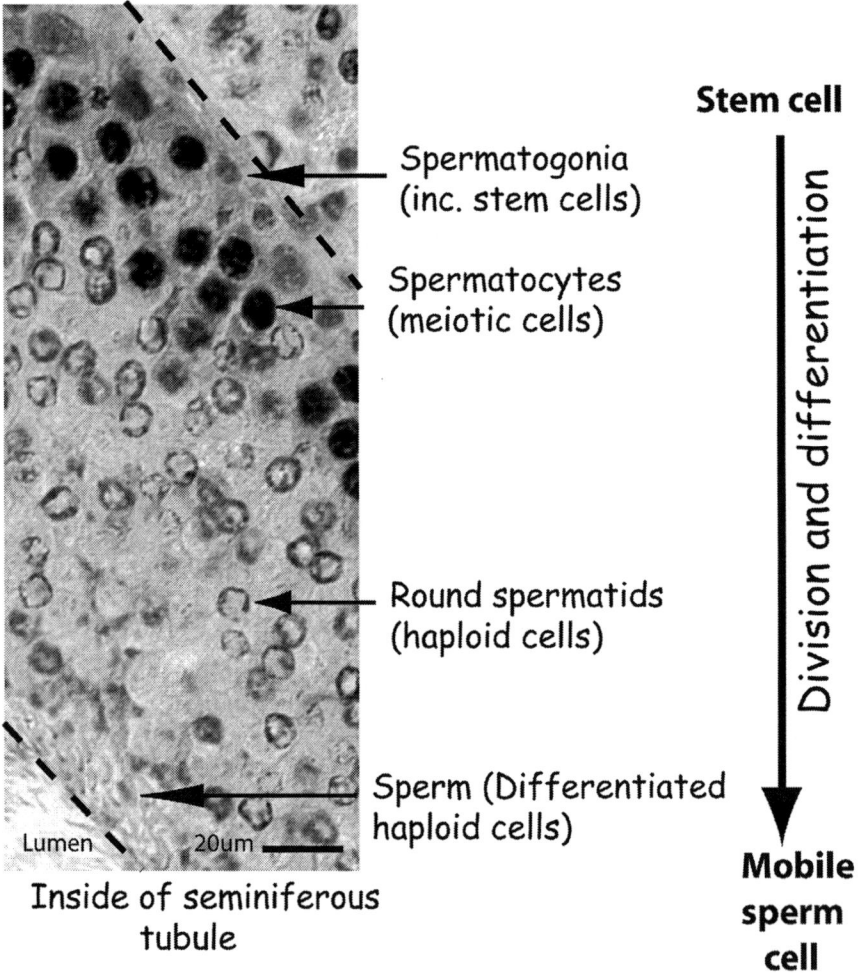

Figure 1. Spermatogenesis. Spermatogenesis is a continuous developmental process which takes place in seminiferous tubules in the adult testis. This micrograph is a section of rat testis stained for Sam68 protein (dark staining in spermatocytes) and counterstained with haematoxylin to show the major germ cell types. Alongside the micrograph the major cell types and direction of differentiation are shown (right hand side: from stem cells at the top of the tissue image, to differentiated sperm at the bottom).

and counterstained using haematoxylin is shown in Figure 1, which shows the anatomic organisation and major stages of male germ cell development.

Germ cell development takes place in the testes within seminiferous tubules (the outside and inside of the tubules are shown as broken lines in Fig. 1). Spermatogenesis is a continuous process but it can be divided up into a number of distinct steps with corresponding cell types.[1] Germ cell development starts from the outside of seminiferous tubules and then progresses towards the inside, with spermatozoa being released into the central hole or lumen of the tubule. The cells at the start of the developmental sequence (and so on the outside of the seminiferous tubule) are called spermatogonia and are mitotically active. Spermatogonial stem cells divide to both renew their own population (this stem cell fraction of spermatogonia are essential to maintain the germline and support future rounds of spermatogenesis) and to provide a source of cells for differentiation (around half of spermatogonia enter the germline differentiation process). A proportion of spermatogonia are eliminated by apoptosis.[2] Differentiating spermatogonia replicate their DNA in an S-phase and then enter meiosis as spermatocytes. The first meiotic division takes around 10 days in mice and during this time homologous chromosomes align, recombine and separate to independently assort alleles. During meiosis I the nuclear envelope remains intact and genes are transcribed. At the end of meiosis I the nuclear envelope breaks down and then sister chromatids separate in meiosis II (in secondary spermatocytes) to generate haploid germ cells. After meiosis the conversion of round spermatids into elongating spermatids involves DNA compaction which shrinks the nucleus of the cell into a compact sperm head, ejection of cytoplasm and growth of a motile flagellum required for sperm locomotion. This differentiation process is called spermiogenesis and creates a small motile cell capable of delivery of genetic information to an egg. The pathway of spermatogenesis (through all the steps from spermatogonia to spermatozoa) takes approximately a month in the mouse and considerably longer in humans.

Distinct genes are required and expressed in the different stages of spermatogenesis depending on the needs of the different cell types. Specific waves of genes are transcribed in mitotic cells, meiotic cells (including genes encoding components of the synaptonemal complex and the meiotic recombination machinery) and in round spermatids (including genes encoding proteins required for more mature sperm).[3] There are also periods of spermatogenesis which are transcriptionally silent, particularly after the haploid genome starts to condense within spermatid nuclei.[4]

As well as transcription of different genes, post-transcriptional control of gene expression plays a critical role in male germ cell development, requiring an important role for RNA binding proteins in the testis.[4] Post-transcriptional gene regulation operates at different levels. Particularly high levels of alternative splicing have been detected in the testis, which is the most abundant anatomic site of alternative splicing apart from the brain. Although many testis-specific splicing isoforms are species specific in pattern, their abundance might reflect an increased demand for mRNA isoform production in the testis because of the ongoing cell division and differentiation.[5]

The mature 3' ends of most mRNAs are generated by an RNA processing event in which nascent pre-mRNA is cleaved by a 3' end processing machinery and a poly(A) tail is added. The 3' end processing machinery recognises a conserved sequence found downstream of genes called the poly(A) site, which includes the sequence AAUAAA. Surprisingly within the testis many meiotic transcripts use variant poly(A) sites and so have alternative 3' ends compared with their somatic counterparts. The testis-expressed protein which recognises these variant poly(A) sites is essential for proper meiotic and postmeiotic development.[6,7]

EXPRESSION OF STAR PROTEINS DURING SPERMATOGENESIS

The STAR proteins are a group of related proteins which contain a KH domain flanked by a pair of conserved sequences called QUA1 and QUA2 domains and a number of other protein domains implicated in cellular signaling pathways.[8] STAR proteins thus have possible roles in both RNA processing and intracellular signaling. Two particular STAR proteins, T-STAR and Sam68, are not only highly expressed in the testis but also interact with other testis-specific RNA binding proteins, suggesting that they may have important roles in RNA processing in mammalian spermatogenesis. T-STAR (also known as SLM2 and KHDRBS3) and Sam68 (also known as KHDRBS1) proteins belong to a subfamily of the STAR proteins called the KH Domain containing RNA Binding Signal transduction associated family. This family of proteins and their encoding genes are consequently given the acronym KHDRBS. The SLM1, Sam68 and T-STAR proteins are encoded by separate genes on mouse chromosomes 1, 4 and 15 respectively.

T-STAR protein is mainly expressed in the adult testis, although weaker protein expression can be detected in the brain and kidney.[9,10] T-STAR is a nuclear protein with major expression in meiotic cells (spermatocytes) and some expression in round spermatids.[10] Sam68 protein is also strongly expressed in spermatocytes (see Fig. 1) and is also expressed outside the testis.[10] The expression of Sam68 protein during spermatogenesis peaks when the meiotic cells get ready to divide.[11] Like T-STAR, Sam68 is also a nuclear protein during most of spermatogenesis, but is found within the cytoplasm in meiosis II.[11,12] Sam68 expression levels decrease in round spermatids when it is again located in the nucleus. Despite its ubiquitous expression in the body, data obtained from the Sam68 null mouse indicates that Sam68 plays an essential role in spermatogenesis rather than being essential for viability.[12,13] The third member of the KHDRBS proteins, SLM-1 (also known as KHDRBS2), is not detectably expressed in the testis.

Published genome sequences predict that the T-STAR, Sam68 and Slm-1 proteins have fairly wide phylogenetic distributions within different vertebrate classes. Orthologous genes with their associated Ensembl identification numbers for mammals (mouse and dolphin), reptiles (Anole lizard), amphibians (frog), fish (zebrafish) and birds (Zebra finch) are shown in Table 1. The mouse and human Sam68 proteins have an N-terminal extension compared with T-STAR and Slm-1 (Fig. 2), as does the dolphin Sam68 protein.

At the moment, there is no genetic evidence or particular expression data suggestive of a specific role for other STAR family proteins in mammalian spermatogenesis. Initial data suggested that the STAR protein Quaking (QKI) may be important for male germ cell development, but in fact this spermatogenesis defect has now been shown to result from co-deletion of the closely linked Parkin co-regulated gene rather than *Qk* itself.[14]

PROTEIN STRUCTURE AND MODIFICATIONS

The protein sequences and domains of human and mouse T-STAR, Sam68 and (for completeness) Slm-1 are shown in Figure 2. Protein homology extends over their entire lengths and can be divided up into four distinct regions with slightly different properties. These distinct regions are: (i) The N-terminal domain, which is considerably longer for Sam68 protein; (ii) The STAR domain, which is composed of the KH domain itself flanked by conserved upstream and downstream regions called QUA1 and QUA2. The STAR domain has roles in both RNA binding and protein interactions including the

Table 1. Phylogenetic distributions of Sam68, T-STAR and Slm-1 genes. Genes were identified on the Ensembl website[58] as cross-species homologues of the mouse STAR genes. Examples are provided for mammals (mouse and dolphin), reptiles (Anole lizard), amphibians (frog), fish (zebrafish) and birds (Zebra finch). The corresponding Ensembl gene identification number is given. Single orthologous genes have been identified in most cases (1 gene copy in each vertebrate species for the *Sam68*, *Slm-1* and *T-STAR* genes), with the exception of 3 predicted gene copies of Sam68 in the zebrafish (i.e., in this case there are 3 zebrafish homologues corresponding to the single mouse Sam68 sequence). The percentage identity of the predicted protein homologue compared with the mouse sequence is shown in the final column. *T-STAR* is also known as *étoile* (French for star), *Slm2* (an acronym of Sam68-like mouse) and *KHDRBS3*. *T-STAR* is encoded by a 9 exon gene on mouse chromosome 8. The T-STAR protein was independently discovered in 3 laboratories. Human T-STAR was initially identified through a yeast 2 hybrid screen which used the male germ cell-specific RNA binding protein RBMY as a bait;[10] independently identified in a similar yeast 2 hybrid screen with rat Scaffold Attachment Factor;[34] and also identified as a protein similar to Sam68 in a database screen and then picked out of a mouse brain library.[59] *Sam68* is also known as *KHDRBS1*. Some entries in column 3 of the table are annotated: *Predicted protein from a partial sequence of a gene which is incomplete at N-terminus or the C-terminus (possibly because the genome has not been completely sequenced); †There is no initial methionine in the predicted protein sequence, but it is otherwise full length.

STAR Family Protein	Other Names	Species and Ensembl ID of Candidate Orthologous Gene	% Identity of Predicted Protein Homologue Over Known Amino Acid Sequence
Sam68	KHDRBS1	Mouse, (Mus musculus) ENSMUSG00000066257	443 amino acids
		Dolphin (Tursiops truncatus) ENSTTRG00000009637	94% over 443 amino acids
		Anole Lizard (Anolis carolinensis) ENSACAG00000011693*	88% over 357 amino acids
		Zebra Finch (Taeniopygia guttata) ENSTGUG00000017346*	93% over 186 amino acids
		Zebrafish (Danio rerio) ENSDARG00000052856,	58% over 370 amino acids
		ENSDARG00000070475, ENSDARG00000078082	64% over 352 amino acids
		Frog (Xenopus tropicalis) ENSXETG00000005246 (pcpb2)	74% over 360 amino acids

continued on next page

Table 1. *continued*

STAR Family Protein	Other Names	Species and Ensembl ID of Candidate Orthologous Gene	% Identity of Predicted Protein Homologue Over Known Amino Acid Sequence
T-STAR	KHDRBS3, étoile, Slm-2	Mouse, (Mus musculus) ENSMUSG00000022332	346 amino acids
		Dolphin (Tursiops truncatus) ENSTTRG00000001270	95% over 346 amino acids
		Anole Lizard (Anolis carolinensis) ENSACAG00000012042[1]	86% over 344 amino acids
		Zebrafish (Danio rerio) Missing- but partial Fugu and stickleback sequences	-
		Zebra Finch (Taeniopygia guttata) ENSTGUG00000012611	85% over 345 amino acids
		Frog (Xenopus tropicalis) ENSXETG00000009227	79% over 344 amino acids
Slm-1	KHDRBS2	Mouse, (*Mus musculus*) ENSMUSG00000026058	349 amino acids
		Dolphin (*Tursiops truncatus*) ENSTTRG00000005591	93% over 348 amino acids
		Anole Lizard (*Anolis carolinensis*) ENSTGUG00000017346*	87% over 322 amino acids
		Zebra Finch (*Taeniopygia guttata*) ENSTGUG00000012786*	89% over 240 amino acids
		Zebrafish (*Danio rerio*) ENSDARG00000069469	74% over 346 amino acids
		Frog (*Xenopus tropicalis*) ENSXETG00000004451	75% over 345 amino acids

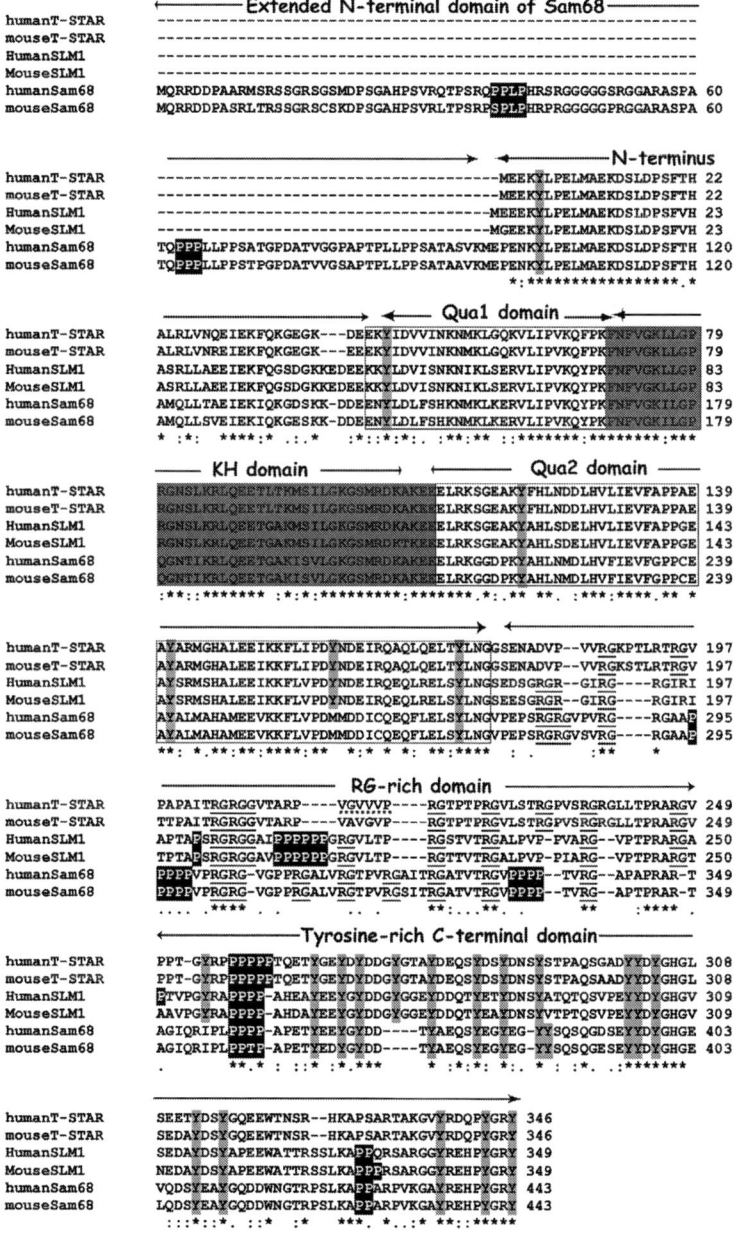

Figure 2. Alignments of the T-STAR, Sam68 and SLM1 protein sequences from human and mouse and the mouse SLM-1 sequence. The KH domain is boxed and shaded grey and the flanking QUA1 and QUA2 domains are boxed only. The region following the QUA2 domain is rich in arginine and glycine and so is referred to as the RG-rich region. RG dipeptides which are potential sites of arginine methylation are underlined. Proline rich sequences that are candidate SH3 binding sites are shaded black and tyrosine residues that might become phosphorylated to become SH2 binding sites are shaded light grey. The C-terminal regions of both T-STAR, Sam68 and SLM1 proteins are particularly rich in tyrosine residues. The mapped SIAH1 binding site in the human T-STAR protein, VGVVVP, is underlined by a broken line.

formation of homodimers and heterodimers with other STAR proteins;[15] (iii) The RG-rich domain, rich in arginine and glycine (RG dipeptides underlined). Some of these arginines are modified by methylation by the enzyme arginine N-methyltransferase 1 (PMRT1). Arginine methylation is important for STAR protein nuclear import and may function to reduce RNA binding;[16,17] (iv) A tyrosine rich C-terminal domain. In addition there are tyrosine residues at equivalent positions throughout the entire proteins (conserved tyrosine residues are shaded light grey in Fig. 2).

Tyrosine residues are potential targets for phosphorylation by protein kinases. In some cases (depending on the sequence context) tyrosine phosphorylation can create SH2 binding sites which act as docking sites for signaling proteins with SH2 domains. Sam68 has been shown to be phosphorylated by a number of tyrosine kinases including Src, Fyn and Lyn[18-21] and is serine/threonine phosphorylated by MAP kinase (also known as ERK) following phorbol ester stimulation of T-cells (Fig. 3).[22] Okadaic Acid (OA) is a phosphatase inhibitor which induces pachytene spermatocytes to enter into meiosis. After in vivo labelling of spermatocytes with ^{32}P in the presence of OA, Sam68 was detectably phosphorylated. Inhibitors of MPF (maturation promoting factor) and MAPKs (mitogen-activated protein kinases) both inhibited the phosphorylation of Sam68 showing these kinases were responsible.[11] A truncated form of the tyrosine kinase receptor (called tr-kit) promotes tyrosine phosphorylation of Sam68 through Fyn in HEK (human embryonic kidney) 293 cells.[23] Unlike the large group of tyrosine kinases which phosphorylate Sam68, the only known protein tyrosine kinase to phosphorylate T-STAR is the epithelial expressed BRK/Sik kinase (also known as Protein Tyrosine Kinase 6). Although BRK/Sik kinase is not detectably expressed in the testis, phosphorylation of transfected T-STAR by transfected BRK/Sik kinase in the mouse mammary gland cell line NMuMG reduced the RNA binding capacity of T-STAR.[24] T-STAR has also been shown to be phosphorylated on tyrosine residues in Rous Sarcoma virus-infected chicken embryo fibroblasts but the kinase responsible is not known.[25]

Both Sam68 and T-STAR proteins also have proline rich SH3-binding sites, which are potential docking sites for proteins that contain SH3 domains (Src homology domain 3). The number and positions of the SH3-binding sites show some conservation between Sam68, T-STAR and Slm1, but are more variable than the tyrosine residues. In particular, Sam68 has 6 potential SH3-binding sites, while T-STAR just has a single SH3-binding site within the RG-rich region (Fig. 2). T-STAR interacts with the p85 subunit of PI3 kinase through its single SH3 binding site.[25] Another domain which can bind to proline rich sequences is known as a WW domain. Sam68 can also bind some WW domain containing proteins, including the formin binding proteins FBP21 and FBP30. Arginine methylation negatively regulates the interactions between Sam68 and the SH3 domain proteins but is permissive for WW domain interactions.[26]

Interaction maps summarising the known protein partners of T-STAR and Sam68 are shown in Figure 3. A clear difference between T-STAR and Sam68 is that human T-STAR protein binds the E3 ubiquitin ligase SIAH1.[9] SIAH1-mediated ubiquitinylation of T-STAR leads to its degradation by the proteasome. The SIAH1 protein is essential for male meiosis in mice,[27] but this is not as a result of regulation of mouse T-STAR protein, since the T-STAR SIAH1 binding site evolved in the primate lineage and so is absent in mice. Hence T-STAR has evolved an extra layer of regulation of protein stability in primates. The mapped SIAH1 binding site within the RG-rich region of human T-STAR protein is underlined by a dashed line in Figure 2. Although the SIAH1 binding site is not conserved in mouse T-STAR, nor within the human Sam68 protein, just two amino substitutions within the mouse T-STAR were sufficient to bring it under the control of SIAH1.[9]

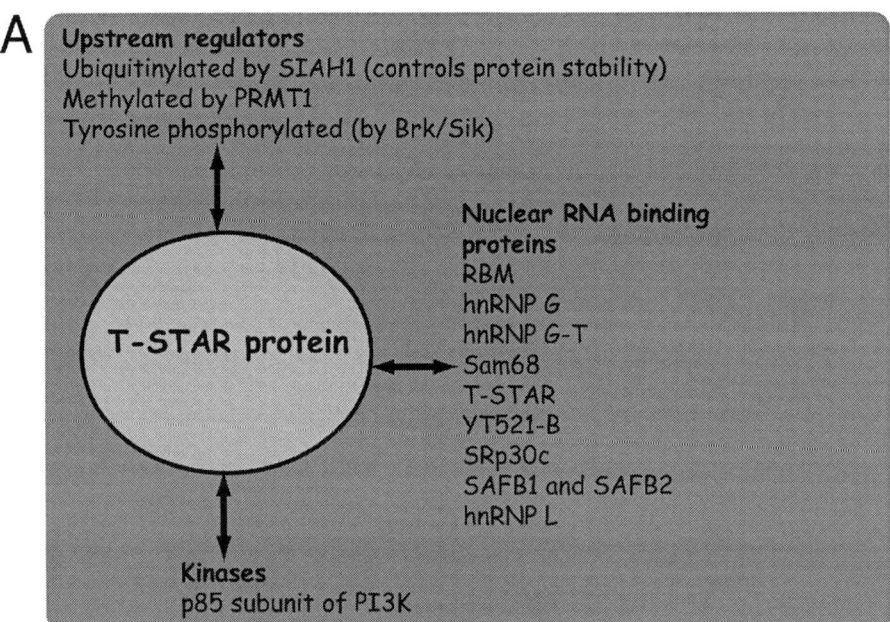

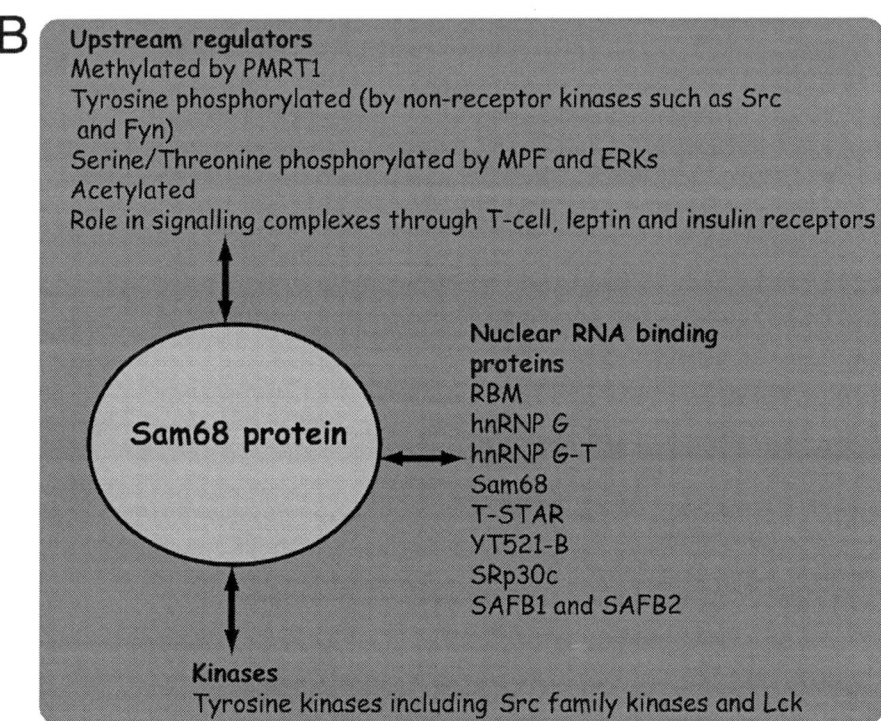

Figure 3. Known protein interaction partners of (A) T-STAR and (B) Sam68. Protein interactions are indicated by double headed arrows and known protein partners are assembled together in groups of interacting proteins which have similar functional properties.

MOUSE KNOCKOUT MODELS DEFINE THE ROLES OF STAR PROTEINS IN TESTIS FUNCTION

Genetic experiments show that Sam68 is essential for male fertility in the mouse.[12,13] Sam68 is encoded by a gene with nine exons. Two of these exons encode part of the KH domain and were deleted to make a *Sam68*[-/-] null mouse. Although *Sam68*[-/-] null females were fertile, *Sam68*[-/-] null male mice were incapable of fathering offspring. When the testis morphology of the knockout and wild type mice were compared histologically, there were fewer round spermatids and elongating spermatids in the *Sam68*[-/-] mice. A significant number of dying cells were also identified in testes by tunnel assay at 25 days post partum in *Sam68*[-/-] mice and flow cytometry indicated that all germ cell stages were affected by an increase in apoptosis. About half of the *Sam68*[-/-] mice made a few sperm, but only about 10% of those made in wild type mice. Amongst the defects observed in the *Sam68*[-/-] mice were sperm with two flagella and two nuclei as well as morphological deformities. The spermatozoa that were made were unable to fertilize eggs by in vitro fertilisation.[12]

THE STAR PROTEIN Sam68 IS INVOLVED IN TRANSLATIONAL CONTROL IN SPERMATOGENESIS

Translational control is particularly important in spermatogenesis and Sam68 protein plays a key role in this process.[4] In the latter stages of spermatogenesis nuclear DNA becomes compacted through replacement of the nuclear histones, first by transition proteins and then by protamines. This means that many mRNAs which are needed in elongating spermatids are in fact transcribed earlier in male germ cell development in the spermatocyte and round spermatid stages and stored in a translationally inactive form until they are needed.

Although it is mainly a nuclear protein in most cell types within the body, during the second meiotic division Sam68 is serine/threonine phosphorylated and translocates to the cytoplasm.[11] When cell extracts from mouse secondary spermatocytes were elutriated and fractionated on sucrose gradients, Sam68 was found to be associated with actively translating polysomes.[12] Sam68 is thought to recruit specific translationally regulated mRNAs to polysomes. In *Sam68*[-/-] null mice these mRNAs are not recruited normally and so become less efficiently translated and thus less abundant. Consistent with this mechanism of translational regulation, in *Sam68*[-/-] null mice compared with heterozygote littermates there was a decrease in the amount of mRNA recruited into polysomes from the three important translationally regulated genes *sperm associated antigen 16, (Spag16), neural precursor cell expressed, developmentally down-regulated gene 1(Nedd1)* and *speedy homolog A (Spdya)*. The Spag16, Nedd1 and Spdya proteins were also reduced in Sam68 null testes, as were levels of their cognate mRNAs. Reduction in translational control may contribute to the infertility phenotype seen in *Sam68*[-/-] null mice. Nedd1 protein is known to localize to the centrosome and mitotic spindle,[28] while Spdya is a member of a family of proteins that binds CDK1 and CDK2 (cyclin dependent protein kinases) and helps oocyte maturation in Xenopus.[29] Spag16 is a component of the sperm axoneme and *Spag16*[-/-] null mice also have defects in sperm motility and are infertile.[30] Although Sam68 clearly has an important role in translation control, to date there is as no evidence to demonstrate an equivalent role for the T-STAR protein in translational control.

STAR PROTEINS MIGHT PLAY ROLES IN PRE-mRNA SPLICING CONTROL IN SPERMATOGENESIS

Although Sam68 is cytoplasmic and involved in translational control in secondary spermatocytes, during most of spermatogenesis both Sam68 and T-STAR proteins are nuclear and likely to be involved in nuclear processes. A possible nuclear role is in alternative splicing regulation. Up to 90% of human genes encode alternatively spliced mRNA isoforms which can have important and divergent functions.[31,32] Sam68 is known to act as a splicing regulator in neural development. Analysis of splicing patterns followed in neuronal cell lines after Sam68 depletion led to the identification of a number of neural alternatively spliced mRNA isoforms which depended on the normal expression of Sam68 protein.[33]

Alternative splicing is controlled in part by networks of nuclear RNA binding proteins which bind to sequences on pre-mRNAs and either serve to activate or repress inclusion of exons. RNA sequences which lead to splicing activation are called splicing enhancers. Splicing enhancers can be found within exons or introns (called Exonic Splicing Enhancers (ESE) and Intronic Splicing Enhancers (ISE) respectively). Similarly cis-acting splicing silencer sites can be found within exons (Exonic Splicing Silencers or ESSs) or within introns (Intronic Splicing Silencers or ISSs).

The mechanisms of splicing activation by Sam68 and T-STAR are not fully understood. Three mechanisms have been proposed and each of these mechanisms might be true for different alternatively spliced exons: (i) Sam68 may operate through binding to ESEs. Co-expression of either Sam68 or T-STAR with a minigene containing exon V5 from the *CD44* gene potently activates splicing of this variable exon.[22,34] Sam68 binds the protein U2AF, which led to the idea that Sam68 bound to an ESE sequence within the *CD44* V5 exon might anchor U2AF binding to the nearby exon V5 3′ splice site and thereby activate its splicing. Within activated T-cells, Sam68 mediated splicing activation of *CD44* exon V5 is enhanced by ERK phosphorylation, which actually leads to release of Sam68 and U2AF proteins from the *CD44* pre-mRNA, perhaps thereby promoting downstream steps in spliceosome assembly and subsequent splicing inclusion of exon V5 into the *CD44* mRNA.[22,35] Sam68 also interacts with the splicing co-activator SRm160 to enhance *CD44* exon V5 splicing, again through facilitating spliceosome assembly.[36] (ii) Sam68 may affect transcriptional elongation. Sam68 protein interacts with the Brahmin protein (also known as Smarca2) which is part of the SWI/SNF complex which regulates chromatin structure. The Sam68-Brahmin protein interaction causes pausing of elongating RNA polymerase around the vicinity of the V5 exon within the chromosomal *CD44* gene. This pause in RNA polymerase II extension would give more chance for the V5 exon, which is weakly recognised by the spliceosome, to be spliced into the *CD44* mRNA before downstream competing exons were transcribed. On the other hand, in the absence of transcriptional pausing through Sam68, stronger downstream exons would be spliced into the *CD44* mRNA instead of the weak exon V5.[37] (iii) Finally Sam68 protein interacts with the splicing repressor protein hnRNP A1 and this interaction might cause selection of proximal versus distal splice sites. This third mechanism regulates BCl-X splice isoform ratios, between mRNAs encoding pro-apoptotic and anti-apoptotic forms of the BCL-X protein.[38]

Although splicing RNA targets of either T-STAR or Sam68 proteins in the testis have not yet been directly identified, protein interactions have been detected between Sam68 and T-STAR and a number of testis-expressed nuclear proteins which have known roles in pre-mRNA splicing.[39] These interacting nuclear proteins include a group of heterogeneous ribonucleoproteins (hnRNPs) including hnRNPG.[34] hnRNP G is encoded by an X-chromosome located gene called *RBMX* which is expressed in spermatogonia and in somatic cells. The hnRNP G protein is known to be involved in controlling pre-mRNA splicing through binding to target RNAs which contain CCA target motifs and also through protein interactions with other splicing factors including the splicing activator Tra2β.[40,41] Within male germ cells Sam68 and T-STAR have two special protein interaction partners related to hnRNP G but encoded by distinct genes: *RBMY*, the Y-chromosome gene deleted in some infertile men[42] and a related protein called hnRNP G-T which is encoded by a retrogene (*RBMXL2*) derived from *RBMX*, but only expressed during male meiosis.[39,43] Normal expression levels of hnRNP G-T protein are needed to support spermatogenesis in mice.[44] Nuclear RNA binding proteins like RBMY and hnRNP G-T in the testis might functionally regulate the splicing activity of T-STAR and Sam68 proteins. Co-expression of hnRNP G with T-STAR inhibits the ability of T-STAR protein to mediate splicing activation of the *CD44* minigene.[34] Protein interactions have also been detected between Sam68 and T-STAR and the splicing regulator proteins SRp30c and YT521B.[34] Other nuclear RNA binding proteins that interact with Sam68 and T-STAR are the Scaffold Attachment Factors SAFB1 and SAFB2 which link signaling, splicing and transcription.[34,45]

OTHER POTENTIAL ROLES OF STAR PROTEINS IN SPERMATOGENESIS

Sam68 has also been implicated in transcriptional regulation.[46] A yeast 2 hybrid screen with human Sam68 against a testis library detected a protein interaction with the tumour suppressor ASPP1,[47] which is also nuclear in human germ cells. ASPP1 increases the selection of apoptosis-promoting target genes by p53.[48]

CONCLUSION

Sam68 and T-STAR proteins interact with RNA via their single KH domains. RNA binding specificities of these proteins have been worked out using SELEX and from sequence characterisation of immunoprecipitated endogenous mRNAs. Recognised target RNA sites in these experiments are AU-enriched sequences such as UAAA.[49-51] Sam68 and T-STAR proteins often require at least 2 AU-rich sequences to bind to synthetic RNAs in SELEX experiments.[49] A key anticipated development is the transcriptome-wide identification of endogenous regulated target RNAs for Sam68 and T-STAR proteins during spermatogenesis. To date there is some tantalizing evidence from immunoprecipitation experiments with the anti-Sam68 antibody in primary spermatocytes:[12] identified target RNAs included *TENR* which encodes a protein related to a family of adenosine deaminases involved in RNA editing,[52] *PP1 gamma*[53] and *Centrin*[54] which encode known mouse spermatogenic proteins.

A disadvantage of traditional immunoprecipitation techniques is that the whole mRNA is recovered, from which it is difficult to deduce the primary binding target sequence. Another possible route to identify endogenous RNA targets at higher resolution is the use of Cross Linking and Immunoprecipitation coupled with high density sequencing (HITS-CLIP), which has been used to identify transcriptome wide targets of the NOVA and argonaute proteins.[55,56] In HITS-CLIP, cells are first UV-irradiated to "freeze" RNA-protein contacts and then trimmed with nucleases, radioactively labelled and immunoprecipitated. HITS-CLIP produces short CLIP tags which correspond to the original binding site of the immunoprecipitated protein and are sequenced at high density. After HITS-CLIP, endogenous RNA targets can be annotated with respect to exons, introns, alternative exons, microRNAs and intergenic transcripts to give a global map of RNA targets and indications of potential roles in RNA processing.

Ideally the identification of endogenous RNA targets for Sam68 and T-STAR should be coupled to the analysis of these RNA targets in wild type and genetically modified mice ablated for either Sam68 or T-STAR proteins. A functional affect on the processing of these target RNAs can then be correlated with the presence or absence of the protein. Identification of specific splicing targets of the T-STAR and Sam68 proteins could also be achieved through the use of microarrays which can detect alternative mRNA isoforms[57] and deep transcriptome sequencing[31,32] in testis RNA from wild type and *Sam68*[-/-] null mice.

Targeting of the gene encoding TSTAR protein has not yet been reported. An anticipated future development is the generation and characterisation of such a mouse strain, which will reveal whether a T-STAR null mouse has a similar or quite different phenotype from the *Sam68*[-/-] null mouse. A further important experiment which will be made possible by the creation of new mouse models will be to test genetically if Sam68 and T-STAR functionally interact in and outside of the germline. For example, is the phenotype of *Sam68*[-/-] null mice more severe when the *KHDRBS3* gene encoding T-STAR is also absent? Linking of mouse genetics and molecular biology should prove a powerful route to understand the biology of STAR proteins in spermatogenesis.

ACKNOWLEDGEMENTS

This work was supported by BBSRC grant BB/D013917/1.

REFERENCES

1. Sassone-Corsi P. Stem cells of the germline: the specialized facets of their differentiation program. Cell Cycle 2008; 7(22):3491-2.
2. Kierszenbaum AL. Apoptosis during spermatogenesis: the thrill of being alive. Mol Reprod Dev 2001; 58(1):1-3.
3. Wrobel G, Primig M. Mammalian male germ cells are fertile ground for expression profiling of sexual reproduction. Reproduction 2005; 129(1):1-7.
4. Paronetto MP, Sette C. Role of RNA-binding proteins in mammalian spermatogenesis. Int J Androl 2009.
5. Kan Z, Garrett-Engele PW, Johnson JM et al. Evolutionarily conserved and diverged alternative splicing events show different expression and functional profiles. Nucleic Acids Res 2005; 33(17):5659-66.
6. Dass B, Tardif S, Park JY et al. Loss of polyadenylation protein tauCstF-64 causes spermatogenic defects and male infertility. Proc Natl Acad Sci USA 2007; 104(51):20374-9.
7. Liu D, Brockman JM, Dass B et al. Systematic variation in mRNA 3'-processing signals during mouse spermatogenesis. Nucleic Acids Res 2007; 35(1):234-46.
8. Vernet C, Artzt K. STAR, a gene family involved in signal transduction and activation of RNA. Trends Genet 1997; 13(12):479-84.
9. Venables JP, Dalgliesh C, Paronetto MP et al. SIAH1 targets the alternative splicing factor T-STAR for degradation by the proteasome. Hum Mol Genet 2004; 13(14):1525-34.
10. Venables JP, Vernet C, Chew SL et al. T-STAR/ETOILE: a novel relative of SAM68 that interacts with an RNA-binding protein implicated in spermatogenesis. Hum Mol Genet 1999; 8(6):959-69.

11. Paronetto MP, Zalfa F, Botti F et al. The nuclear RNA-binding protein Sam68 translocates to the cytoplasm and associates with the polysomes in mouse spermatocytes. Mol Biol Cell 2006; 17(1):14-24.

12. Paronetto MP, Messina V, Bianchi E et al. Sam68 regulates translation of target mRNAs in male germ cells, necessary for mouse spermatogenesis. J Cell Biol 2009; 185(2):235-49.

13. Richard S, Torabi N, Franco GV et al. Ablation of the Sam68 RNA binding protein protects mice from age-related bone loss. PLoS Genetics 2005; 1(6):e74.

14. Lorenzetti D, Bishop CE, Justice MJ. Deletion of the Parkin co-regulated gene causes male sterility in the quaking(viable) mouse mutant. Proc Natl Acad Sci USA 2004; 101(22):8402-7.

15. Chen T, Damaj BB, Herrera C et al. Self-association of the single-KH-domain family members Sam68, GRP33, GLD-1 and Qk1: role of the KH domain. Mol Cell Biol 1997; 17(10):5707-18.

16. Cote J, Boisvert FM, Boulanger MC et al. Sam68 RNA binding protein is an in vivo substrate for protein arginine N-methyltransferase 1. Mol Biol Cell 2003; 14(1):274-87.

17. Rho J, Choi S, Jung CR et al. Arginine methylation of Sam68 and SLM proteins negatively regulates their poly(U) RNA binding activity. Arch Biochem Biophys 2007; 466(1):49-57.

18. Fumagalli S, Totty NF, Hsuan JJ et al. A target for Src in mitosis. Nature 1994; 368(6474):871-4.

19. Richard S, Yu D, Blumer KJ et al. Association of p62, a multifunctional SH2- and SH3-domain-binding protein, with src family tyrosine kinases, Grb2 and phospholipase C gamma-1. Mol Cell Biol 1995; 15(1):186-97.

20. Taylor SJ, Shalloway D. An RNA-binding protein associated with Src through its SH2 and SH3 domains in mitosis. Nature 1994; 368(6474):867-71.

21. Weng Z, Thomas SM, Rickles RJ et al. Identification of Src, Fyn and Lyn SH3-binding proteins: implications for a function of SH3 domains. Mol Cell Biol 1994; 14(7):4509-21.

22. Matter N, Herrlich P, Konig H. Signal-dependent regulation of splicing via phosphorylation of Sam68. Nature 2002; 420(6916):691-5.

23. Paronetto MP, Venables JP, Elliott DJ et al. Tr-kit promotes the formation of a multimolecular complex composed by Fyn, PLCgamma1 and Sam68. Oncogene 2003; 22(54):8707-15.

24. Haegebarth A, Heap D, Bie W et al. The nuclear tyrosine kinase BRK/Sik phosphorylates and inhibits the RNA-binding activities of the Sam68-like mammalian proteins SLM-1 and SLM-2. J Biol Chem 2004; 279(52):54398-404.

25. Lee J, Burr JG. Salpalpha and Salpbeta, growth-arresting homologs of Sam68. Gene 1999; 240(1):133-47.

26. Bedford MT, Frankel A, Yaffe MB et al. Arginine methylation inhibits the binding of proline-rich ligands to Src homology 3, but not WW, domains. J Biol Chem 2000; 275(21):16030-6.

27. Dickins RA, Frew IJ, House CM et al. The ubiquitin ligase component Siah1a is required for completion of meiosis I in male mice. Mol Cell Biol 2002; 22(7):2294-303.

28. Manning J, Kumar S. NEDD1: function in microtubule nucleation, spindle assembly and beyond. Int J Biochem Cell Biol 2007; 39(1):7-11.

29. Gastwirt RF, McAndrew CW, Donoghue DJ. Speedy/RINGO regulation of CDKs in cell cycle, checkpoint activation and apoptosis. Cell Cycle 2007; 6(10):1188-93.

30. Zhang Z, Kostetskii I, Tang W et al. Deficiency of SPAG16L causes male infertility associated with impaired sperm motility. Biol Reprod 2006; 74(4):751-9.

31. Pan Q, Shai O, Lee LJ et al. Deep surveying of alternative splicing complexity in the human transcriptome by high-throughput sequencing. Nat Genet 2008; 40(12):1413-5.

32. Wang ET, Sandberg R, Luo S et al. Alternative isoform regulation in human tissue transcriptomes. Nature 2008; 456(7221):470-6.

33. Chawla G, Lin CH, Han A et al. Sam68 regulates a set of alternatively spliced exons during neurogenesis. Mol Cell Biol 2009; 29(1):201-13.

34. Stoss O, Olbrich M, Hartmann AM et al. The STAR/GSG family protein rSLM-2 regulates the selection of alternative splice sites. J Biol Chem 2001; 276(12):8665-73.

35. Tisserant A, Konig H. Signal-regulated Pre-mRNA occupancy by the general splicing factor U2AF. PloS One 2008; 3(1):e1418.

36. Cheng C, Sharp PA. Regulation of CD44 alternative splicing by SRm160 and its potential role in tumor cell invasion. Mol Cell Biol 2006; 26(1):362-70.

37. Batsche E, Yaniv M, Muchardt C. The human SWI/SNF subunit Brm is a regulator of alternative splicing. Nat Struct Mol Biol 2006; 13(1):22-9.

38. Paronetto MP, Achsel T, Massiello A et al. The RNA-binding protein Sam68 modulates the alternative splicing of Bcl-x. J Cell Biol 2007; 176(7):929-39.

39. Venables JP, Elliott DJ, Makarova OV et al. RBMY, a probable human spermatogenesis factor and other hnRNP G proteins interact with Tra2beta and affect splicing. Hum Mol Genet 2000; 9(5):685-94.

40. Heinrich B, Zhang Z, Raitskin O et al. Heterogeneous nuclear ribonucleoprotein G regulates splice site selection by binding to CC(A/C)-rich regions in pre-mRNA. J Biol Chem 2009; 284(21):14303-15.

41. Liu Y, Bourgeois CF, Pang S et al. The germ cell nuclear proteins hnRNP G-T and RBMY activate a testis-specific exon. PLoS Genetics 2009; 5(11):e1000707.

42. Elliott DJ, Millar MR, Oghene K et al. Expression of RBM in the nuclei of human germ cells is dependent on a critical region of the Y chromosome long arm. Proc Natl Acad Sci USA 1997; 94(8):3848-53.

43. Elliott DJ, Venables JP, Newton CS et al. An evolutionarily conserved germ cell-specific hnRNP is encoded by a retrotransposed gene. Hum Mol Genet 2000; 9(14):2117-24.

44. Ehrmann I, Dalgliesh C, Tsaousi A et al. Haploinsufficiency of the germ cell-specific nuclear RNA binding protein hnRNP G-T prevents functional spermatogenesis in the mouse. Hum Mol Genet 2008; 17(18):2803–18.

45. Sergeant KA, Bourgeois CF, Dalgliesh C et al. Alternative RNA splicing complexes containing the scaffold attachment factor SAFB2. J Cell Sci 2007; 120(Pt 2):309-19.

46. Lukong KE, Richard S. Sam68, the KH domain-containing superSTAR. Biochim Biophys Acta 2003; 1653(2):73-86.

47. Thornton JK, Dalgliesh C, Venables JP et al. The tumour-suppressor protein ASPP1 is nuclear in human germ cells and can modulate ratios of CD44 exon V5 spliced isoforms in vivo. Oncogene 2006; 25(22):3104-12.

48. Sullivan A, Lu X. ASPP: a new family of oncogenes and tumour suppressor genes. Br J Cancer 2007; 96(2):196-200.

49. Galarneau A, Richard S. The STAR RNA binding proteins GLD-1, QKI, SAM68 and SLM-2 bind bipartite RNA motifs. BMC Mol Biol 2009; 10:47.

50. Lin Q, Taylor SJ, Shalloway D. Specificity and determinants of Sam68 RNA binding. Implications for the biological function of K homology domains. J Biol Chem 1997; 272(43):27274-80.

51. Tremblay GA, Richard S. mRNAs associated with the Sam68 RNA binding protein. RNA Biol 2006; 3(2):90-3.

52. Connolly CM, Dearth AT, Braun RE. Disruption of murine Tenr results in teratospermia and male infertility. Dev Biol 2005; 278(1):13-21.

53. Varmuza S, Jurisicova A, Okano K et al. Spermiogenesis is impaired in mice bearing a targeted mutation in the protein phosphatase 1cgamma gene. Dev Biol 1999; 205(1):98-110.

54. Klink VP, Wolniak SM. Centrin is necessary for the formation of the motile apparatus in spermatids of Marsilea. Mol Biol Cell 2001; 12(3):761-76.

55. Chi SW, Zang JB, Mele A et al. Argonaute HITS-CLIP decodes microRNA-mRNA interaction maps. Nature 2009.

56. Licatalosi DD, Mele A, Fak JJ et al. HITS-CLIP yields genome-wide insights into brain alternative RNA processing. Nature 2008; 456(7221):464-9.

57. Ule J, Ule A, Spencer J et al. Nova regulates brain-specific splicing to shape the synapse. Nat Genet 2005; 37(8):844-52.

58. Hubbard TJ, Aken BL, Ayling S et al. Ensembl 2009. Nucleic Acids Res 2009; 37(Database issue):D690-7.

59. Di Fruscio M, Chen T, Richard S. Characterization of Sam68-like mammalian proteins SLM-1 and SLM-2: SLM-1 is a Src substrate during mitosis. Proc Nat Acad Sci USA 1999; 96(6):2710-5.

CHAPTER 6

THE ROLE OF QUAKING IN MAMMALIAN EMBRYONIC DEVELOPMENT

Monica J. Justice* and Karen K. Hirschi

Abstract: Functional studies of the mouse *quaking* gene (*Qk*) have focused on its role in the postnatal central nervous system during myelination. However, the death of the majority of homozygous mouse *quaking* alleles revealed that *quaking* has a critical role in embryonic development prior to the start of myelination. Surprisingly, the lethal alleles revealed that *quaking* has a function in embryonic blood vessel formation and remodeling. Further studies of the extraembryonic yolk sac showed that *Qk* regulates visceral endoderm differentiated function at the cellular level, including the local synthesis of retinoic acid (RA), which then exerts paracrine control of endothelial cells within adjacent mesoderm. Endoderm-derived RA regulates proliferation of endothelial cells and extracellular matrix (ECM) production, which in a reciprocal manner, modulates visceral endoderm survival and function. Although exogenous RA can rescue endothelial cell growth control and ECM production in mutants carrying a lethal mutation, which lack functional *Qk*, neither visceral endoderm function nor vascular remodeling is restored. Thus, *Qk* also regulates cell autonomous functions of visceral endoderm that are critical for vascular remodeling. Interestingly, *quaking* is highly expressed during normal cardiac development, particularly in the outflow tract, suggesting potentially unique functions in the developing heart. Together, the work on *Qk* in mammalian embryos reveals an essential, yet under appreciated, role in cardiovascular development. This suggests that certain functions may remain conserved in the early embryo throughout the evolution of nonvertebrate and vertebrate organisms and that additional roles for *quaking* remain to be discovered.

*Corresponding Author: Monica J. Justice—Departments of Molecular and Human Genetics and Molecular Physiology and Biophysics, Baylor College of Medicine, One Baylor Plaza MS227, R804, Houston, Texas 77030, USA. Email: mjustice@bcm.edu

Post-Transcriptional Regulation by STAR Proteins: Control of RNA Metabolism in Development and Disease, edited by Talila Volk and Karen Artzt.

INTRODUCTION

In the mouse, the *quaking* (*Qk*) gene, a member of the STAR family of signal transduction and RNA binding proteins, was originally defined for its function in central nervous system myelination; however, *Qk* also functions in early vascular development.[1,2] Three protein isoforms of QK (QKI5, QKI6 and QKI7), produced by alternative splicing, share a KH domain, but differ in their carboxy-termini.[3] QKI5, which contains a nuclear localization signal in its carboxy-terminus, regulates alternative splicing in vitro[4,5] and may regulate the splicing of *Qk* itself[1] (MJJ unpublished). QKI6 and QKI7 are predominantly cytoplasmic and QKI6 is essential for central nervous system myelination.[4,6,7]

The multiple in vivo functions of *Qk* have been revealed by an allelic series of mutant mice. The *quaking viable* (*Qk^v*) recessive allele, a spontaneously occurring 1 Mb deletion within the promoter/enhancer region of *Qk*[8] affects oligodendrocytes and Schwann cells resulting in hypomyelination.[6] This mutation also deletes the Parkin Co-regulated gene (*Pacrg*) and the gene responsible for juvenile autosomal recessive Parkinsons's disease (*Parkin*).[9,10] A series of *N*-ethyl-*N*-nitrosourea (ENU)-induced point mutations, *Qk^{l-1}*, *Qk^{kt1}*, *Qk^{kt3/4}* and *Qk^{k2}*, as well as a knockout deletion allele, have demonstrated a role for *Qk* independent of myelination, as these alleles when homozygous cause embryonic death at midgestation due to cardiovascular failure (Table 1).[11-13]

The *Qk^{kt3/4}* allele, which is an A to G transition, results in a glutamic acid to glycine change in the QUA1 region that eliminates protein dimerization, creating a null allele.[11,12,14] The *Qk^{k2}* mutation is a T to A transversion which changes a valine to a glutamic acid

Table 1. Mouse *quaking* alleles

Allele*	Homozygous Phenotype (Heterozygous Phenotype)	Lesion
Qk^v	Quaking, males sterile	1 Mb deletion flanking *Qk*, which alters *Qk* expression and deletes *Parkin* and *Pacrg*
Qk^{kt1}	Embryonic lethal (Spontaneous seizures in aged heterozygotes)	Unknown, not in coding sequence
Qk^{k2}	Embryo lethal with vascular insufficiency and heart defects (Semidominant reduction of adult brain myelin lipids, susceptibility to seizures)	V to G change in KH domain
Qk^{kt3/4}	Embryo lethal with vascular insufficiency and heart defects	E to G change/loss of protein dimerization
Qk^{l-1}	Embryo lethal with vascular in-suf-ficiency and heart defects (Transient quaking in compound heterozygotes *Qk^{l-1}/Qk^v*)	Splice defect which results in loss of QKI5 isoform
Qk^{tm1Abe}	Embryonic lethal with vascular insufficiency	Deletion of Exon 1, including the translational start site
Qk^{e5}	Extremely severe quaking and seizures Male fertile	Unknown, not in coding sequence,[76] maps between 40 kb and 640 kb upstream of gene

in the KH domain.[2,11] The Qk^{l-1} mutation abolishes the splice site necessary to produce the $QkI5$ transcript by an A to G transition and homozygous Qk^{l-1}/Qk^{l-1} mutants die at midgestation of vascular insufficiency.[1,4,11,15] The mutation responsible for the Qk^{kt1} embryonic lethal allele has yet to be identified.[11] The Qk^{l-1} and Qk^{k2} alleles, along with the knockout allele, were used to show that defective Qk function leads to a failure of vascular remodeling resulting in embryonic death.[2]

QUAKING IS REQUIRED FOR THE FORMATION OF EMBRYONIC VASCULATURE

QKI regulates vascular development during embryogenesis.[2] Mammalian blood vessel formation begins via a process termed vasculogenesis, or the de novo differentiation of endothelial cells. It initiates shortly after gastrulation in the yolk sac mesoderm, which lies adjacent to visceral (primitive) endoderm (Fig. 1A).[16] Mesodermal progenitors are thought to receive cues from the visceral endoderm to direct their differentiation into primitive endothelial and hematopoietic cells, which collectively constitute blood islands.[17,18] Later stages of vasculogenesis include formation of vascular channels and a capillary plexus, which is then remodeled via a process referred to as angiogenesis, into a circulatory network composed of specialized endothelial cell types (i.e., arterial, venous).[19,20] During vascular remodeling, endothelial cell tubes are stabilized via recruitment of and cell–cell interactions with, surrounding mural cells (vascular smooth muscle cells and pericytes) and extracellular matrix produced by both cell types (Fig. 1A).

By embryonic day (E) 9.5, wild-type yolk sacs in the mouse have a well-developed vascular system composed of endothelial cell tubes invested by surrounding mural cells. Mice homozygous for the Qk^{k2} or Qk^{l-1} alleles have a primary defect in yolk sac and embryonic vascular remodeling prior to the recruitment of smooth muscle cells, resulting in vascular insufficiency and embryonic lethality by E10.5. A similar vascular insufficiency phenotype is found in the knockout allele, although the conclusions drawn about the primary defect are somewhat different, suggesting that quaking directly controls smooth muscle cell recruitment and/or differentiation.[13] Importantly, the QKI5 isoform is not expressed in endothelial or smooth muscle cells that form the vasculature, but rather is expressed in the endoderm adjacent to vascular cells in the yolk sac at E8.5 and 9.5 (Fig. 2A); thus, a cell autonomous role in endothelial and/or mural cell development for QKI5 is not likely.

QKI5 REGULATES $QKI6$ AND $QKI7$ IN VISCERAL ENDODERM

The $quaking$ allelic series is a valuable tool to determine the cellular and molecular roles of Qk in vascular development. Previous studies suggested that QKI5 was found mainly in embryos and that QKI6 and QKI7 were found in oligodendrocytes. However, more detailed investigation showed that all three isoforms are found in the visceral endoderm of the extraembryonic yolk sac.[1] Using two Qk alleles, Qk^{k2}, which has a point mutation in the KH domain to produce a mutant QKI5 isoform and Qk^{l-1}, which completely lacks the QKI5 isoform, we demonstrated that QKI5, through its KH-domain, regulates the expression of the $QkI6$ and $QkI7$ transcripts. In both $Qk^{k2}/$

Qk^{k2} and Qk^{l-1}/Qk^{l-1} mutant yolk sacs, the expression of the $QkI6$ and $QkI7$ transcripts was decreased, suggesting that functional QKI5 is required for vascular development and could regulate the expression of the other two Qk isoforms through alternative splicing or stabilization. Both mutants fail to remodel the capillary plexus in the embryo and yolk sac.

QKI5 regulates alternative splicing of proteins associated with myelination in adult brain.[5] In Qk^v/Qk^v mutants, decreased levels of QKI5 expression correlates with increased severity of dysmyelination and the ratio of alternative splice variants of several myelin genes is altered. The QKI protein binds a consensus hexonucleotide (UACU(C/A)A) found in the 3'UTR of myelin basic protein (MBP) mRNA transcripts.[21] This RNA sequence was first identified using the closely related *Caenorhabditis elegans* STAR-protein germline defective (GLD-1) and allows transcripts containing it to bind to the conserved KH and QUA2 domains.[21-24] Interestingly, this conserved hexonucleotide sequence lies in the unspliced Qk transcript and the QKI5 isoform lending support to the idea that QKI binds its own transcript to regulate expression of the alternate $QkI6$ and $QkI7$ transcripts[1](Unpublished observations, MJJ).

MOLECULAR BASIS OF BLOOD VESSEL FORMATION

Vasculogenesis and angiogenesis require the coordination of a complex orchestration of cellular differentiation, proliferation and migration and a multitude of different growth factors and signal transduction pathways. Each component functions to regulate distinct stages of blood vessel assembly (reviewed in refs. 17,25-27). Our present knowledge of the regulation of mammalian endothelial cell differentiation and its specialization has been largely derived from studies of mouse embryonic development. In this model system, it is suggested that during yolk sac vasculogenesis, visceral endoderm-derived soluble factors, such as Indian Hedgehog (IHH),[28,29] vascular endothelial growth factor (VEGF)[30,31] and basic fibroblast growth factor (bFGF)[32,33] promote endothelial cell formation within the underlying mesoderm where their receptors, Patched (Ptc),[34] VEGF receptor 2 (VEGFR2/Flk1)[35,36] and fibroblast growth factor receptor 2 (FGFR2),[37] respectively, are localized (reviewed in refs. 26,38,39). Other signaling molecules proposed to be downstream targets of IHH and VEGF signaling, such as bone morphogenetic protein 4 (BMP4), are similarly localized within the mesoderm.[29,40] While all of these factors are individually important in regulating endothelial cell formation, the signaling hierarchy among them has not been clearly delineated in vivo.

Vascular remodeling involves other signaling pathways including retinoic acid (RA),[41] which directly regulates endothelial cell proliferation and indirectly regulates endothelial cell migration and visceral endoderm survival.[42] Angiopoietin 1 (Ang1), its antagonist Ang2 and the endothelial-specific receptor tyrosine kinase (Tek/Tie2), also regulate capillary remodeling and maturation of the endothelial cells, enabling them to recruit mural cell progenitors[43,44] in a process modulated by Platelet-derived growth factor-beta (PDGFB) signaling.[45,46] Differentiation of mural cell progenitors occurs via transforming growth factor beta (TGFB) signaling[47] and requires direct endothelial—mural cell progenitor contact and gap junction channel formation.[48] Other signal transduction factors and their receptors are also critical for early blood vessel formation and remodeling including Jagged1 and several forms of Notch.[49-51] In addition, many other factors are essential for proper vascularization including transcription factors T-cell acute lymphocytic leukemia 1 (Tal1/SCL) and nuclear receptor subfamily 2, group F, member 2 (Nr2f2/COUP-TFII), extracellular matrix proteins laminin and fibronectin (Fn) and cell adhesion molecules that include platelet endothelial cell adhesion molecule 1 (PECAM1), cadherin 5 (Cdh5/VE-CAD) and intercellular adhesion molecule 1 (ICAM1) (reviewed in ref. 52). Not surprisingly then, vascular insufficiency is a common phenotype shared by numerous genetically-altered mice with mutations in some of these pathways including vascular endothelial growth factor A (VEGFA), IHH, bFGF, TGFB, Fn, Notch1, Mindbomb1 (*Mib1*) and RA[30,33,41,53-57] (reviewed in ref. 27).

QUAKING IS REQUIRED FOR VISCERAL ENDODERM DIFFERENTIATED FUNCTION

Although the molecular function of QKI in the early embryo is still under investigation, we have determined that an essential cellular function of QKI is the regulation of visceral endoderm differentiated function (Fig. 1B). A mature visceral endoderm is needed for production of junctional proteins (i.e., claudin 7 (Cldn7), serum proteins (i.e., alpha fetoprotein (AFP)), metabolic enzymes (i.e., aldehyde dehydrogenase family 1, subfamily 2 (Aldh1a2/Raldh2)) and soluble growth factors (i.e., VEGF-A, IHH, bFGF) that modulate vascular development (Fig. 1A).[52,58] Interestingly, Aldh1a2/Raldh2, the enzyme required for RA synthesis, was downregulated in Qk^{k2}/Qk^{k2} and Qk^{l-1}/Qk^{l-1} mutants.[1] We previously demonstrated that RA targets endothelial cells in the mesoderm via retinoic acid receptors alpha 1 and 2 (RARα1/2) specifically expressed therein[42] and a hierarchy of signals downstream of RA is necessary for endothelial cell growth control and migration, as well as visceral endoderm survival.[42] Similar to RA-deficient mutants, Qk^{k2}/Qk^{k2} mutants exhibited increased endothelial cell proliferation and visceral endoderm apoptosis (Fig. 1B). Importantly, these specific defects in *Qk* mutant embryos can be rescued by in vivo RA feedings of pregnant females. The RA feedings also restored expression of VEGF1, IHH and bFGF. Although QKI is an intracellular protein, its regulation of RA synthesis via Raldh2 production also mediates developmental processes in the adjacent mesoderm where RARs are expressed in endothelial cells. That is, a defect in QKI protein function leads to secondary RA deficiency, which causes lack of endothelial cell growth control and maturation and subsequent visceral endoderm apoptosis. RA did not, however, restore all defects such as serum and tight junction protein production (AFP, Cldn7 and Raldh2) (Fig. 1B). Further, vascular remodeling and embryonic survival were not rescued in spite of the integrity of the visceral endoderm being restored. Therefore, QKI likely has direct targets within the visceral endoderm, independent of RA-mediated signaling in the mesoderm, that are required for proper vascular patterning and remodeling. Together these data show that the *Qk* gene modulates vascular development via its regulation of visceral endoderm differentiated function. These findings highlight the importance of visceral endodermal-mesodermal interactions and hierarchies among signaling pathways that direct vascular remodeling and embryonic morphogenesis.

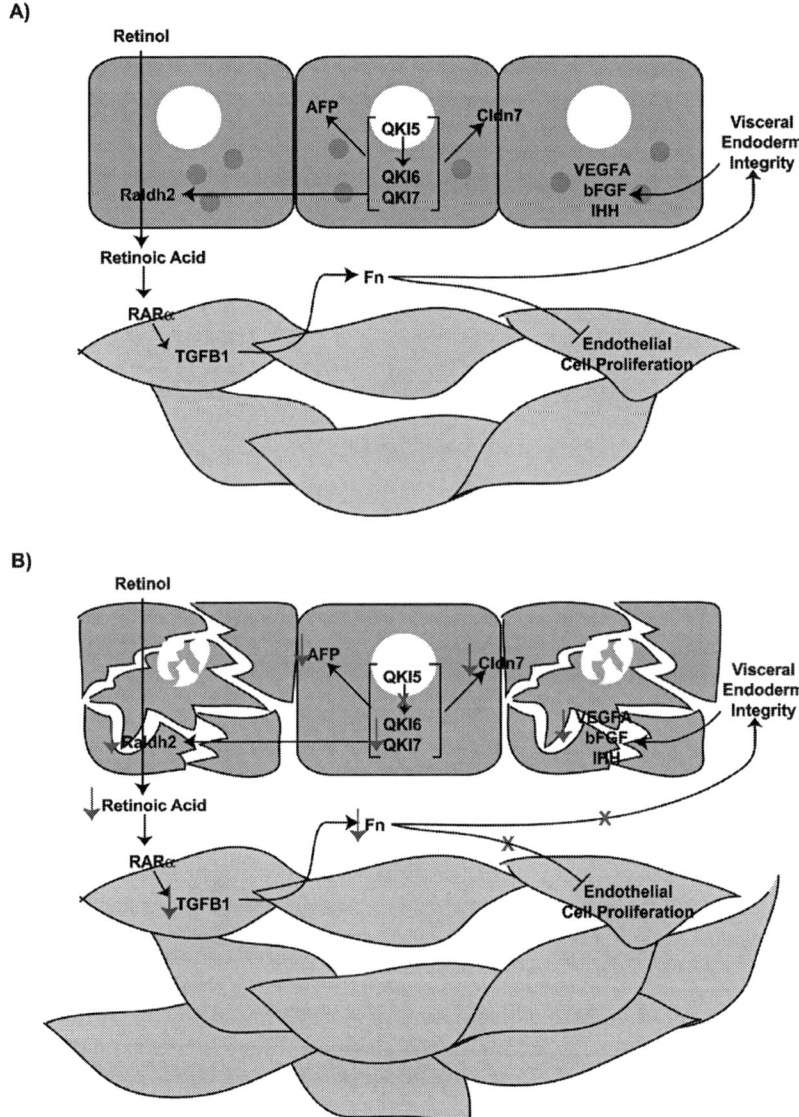

Figure 1. Model of the role of quaking in the vascular endothelium. A) In wild type yolk sacs, QKI5 regulates expression of QKI6 and QKI7 in the visceral endoderm. Production of serum proteins (AFP) and metabolic enzymes (Aldh1a2/Raldh2), as well as formation of an epithelial barrier (Cldn7) is dependent on QKI. Retinoic acid, produced by the visceral endoderm, targets endothelial cells within the mesoderm via RARa and regulates TFGB1-mediated Fn deposition. Fn is necessary for growth factor production (VEGFA, IHH and bFGF). Figure is adapted from.[1] B) In homozygous quaking mutant yolk sacs, mutation of the *Qk* gene results in decreased expression of QKI6 and QKI7. Visceral endoderm differentiated function is compromised, leading to decreased retinoic acid signaling and subsequent loss of Fn deposition. These alterations result in lack of endothelial growth control and decreased visceral endoderm survival. Production of visceral endoderm growth factors (VEGFA, IHH and bFGF) are mediated by retinoic acid-induced Fn, however, other functions of the visceral endoderm are directly regulated by QKI and independent of retinoic acid.

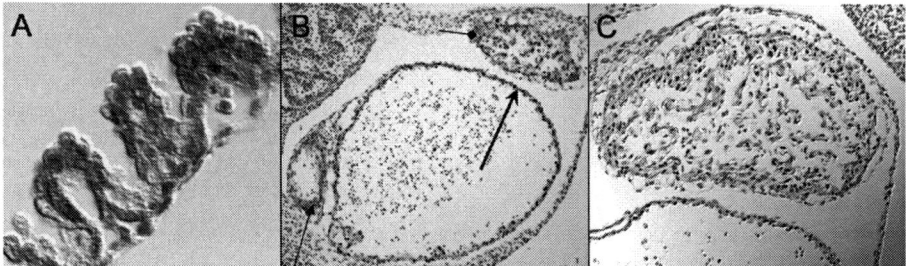

Figure 2. QKI5 localization in E10.5 wild type embryos. A) Expression of QKI5 protein as detected by immunohistochemistry is localized to the endodermal layer of the yolk sac, adjacent to differentiated endothelial cells in the mesodermal layer. B) QKI5 protein in the endocardium of the common atrium (large arrow) and outflow tract (diamond arrow, this is the Bulbis Cordis in humans), as well as the sinus venosus (filled block arrow). C) Magnified view of the endocardium of the common atrium and the outflow tract.

OTHER POSSIBLE ROLES FOR QUAKING IN CARDIOVASCULAR DEVELOPMENT

It is clear that the cause of death in *quaking* lethal mouse embryos results from its earliest function in extraembryonic visceral endoderm. However, we expect that other embryonic functions for quaking remain to be identified. For example, although we did not find expression of QKI5 in embryonic mural cells, a reporter gene in the knockout allele suggests that at least some isoform of quaking is expressed in these cells.[13] The Qk^{k2}/Qk^{k2} and the Qk^{l-1}/Qk^{l-1} mutants exhibit decreased vascular remodeling in the embryo proper, the outcome of which is severe hemorrhaging.[2,11,12] Further, *quaking* embryos have morphological heart defects, especially in the outflow tract; however, we previously demonstrated that in the Qk^{k2}/Qk^{k2} mutants, cardiac differentiation and myocardial function is not compromised at E9.5.[2,12,13] Localization of the protein using isoform-specific antibodies shows that QKI5 is expressed in the endocardium of the common atrium, outflow tract and sinus venosus of the developing heart at E10.5 (Fig. 2B,C). The outflow tract, which consists of neural crest and mesoderm derivatives, will be remodeled into the aorta and pulmonary trunk. Together these data suggest that *Qk* is likely required for both extraembryonic vascular and embryonic cardiovascular development. Tissue-specific conditional knockout alleles of *quaking* will be required to differentiate its roles in embryonic and extraembryonic vasculature and primary cardiac function.

THE EVOLVING ROLES OF QUAKING FUNCTION

The *quaking* gene is highly conserved in human (*QKI*), Xenopus (*Xqua*), chicken (*qk*) and Drosophila (*who/how/qkr93F*). The expression patterns of *Qk* homologues in vertebrate and nonvertebrate species and the loss of function phenotypes generated in Drosophila, Xenopus and mouse imply that the earliest vertebrate function of quaking is in the development of the heart and vascular tissues.[1,59-64] Tanaka et al[65] proposed that the primordial function of quaking originated in the mesoderm to regulate cardiac and muscle development as seen in Drosophila and that in vertebrates an additional function evolved in the nervous system. However, the role of Qk in developing glial progenitors is also likely to be functionally conserved.[66] In fact, vasculogenesis and neurogenesis are coupled during embryonic development to create an intricately linked branching pattern of nerves and vessels.[67,68] VEGF signaling is important for blood vessel formation and VEGF stimulates axonal growth and cell survival.[69] Several gene families and signaling systems long studied for their roles in neural development and patterning are also required for vascular development, including Notch, Delta, ephrins, Slit and Roundabout (robo) (reviewed in refs. 67,70). Therefore, clues to quaking's function in vertebrate cardiovascular development may be found in its role in flies (see Chapter 7 by T. Volk). The development of the cardiovascular system in Drosophila has been compared with the formation of primary vessels in vertebrates.[71] Hypomorphic alleles of the Drosophila homologue (*How*) have abnormal wing musculature resulting in the phenotype of held out wings; however, loss of function alleles die prior to hatching with lack of muscle movement, narrowed dorsal aortas and low heart rate.[64] Recent analyses of the signaling pathways involved in the morphogenesis of the *Drosophila* cardiac tube show that *How* acts upstream of *slit* and *robo* and interacts genetically with them.[72] Slit is a secreted glycoprotein that signals through its receptor robo to function in axon guidance and blood vessel patterning.[73-75] Perhaps quaking will also function upstream of vertebrate slits in the induction of vessels throughout the embryo, as well as in the formation of heart structures such as the outflow tract. By studying pathways conserved in the vascular and nervous systems in multiple model organisms, we may learn whether one system was altered during the process of evolution to adapt to a new system based on the same signaling mechanisms.

CONCLUSION

Our work on quaking has raised several interesting questions for further studies of the role of quaking in mammalian embryonic development. Quaking plays a role in the extraembryonic visceral endoderm, which leads to early embryonic vascular failure and death, possibly masking later acting roles. The expression of quaking isoforms in the embryo proper suggests additional roles in mesodermal, neuroepithelial and neural crest cells. Does quaking have an autonomous function in the developing mammalian heart? What myriad of distinct functions may the three isoforms perform in different cell types? Additional studies are needed to examine expression in cells of the early embryo and developing heart. Are signaling mechanisms used by quaking in mammalian embryonic blood vessel formation and/or the mammalian heart similar to those in Drosophila heart morphogenesis? Qk may play a role in the induction, patterning and development of vessels in the embryo proper and this role may be linked to the slit robo pathway. Tools such as mouse conditional mutations and cardiovascular imaging systems to image live embryos are now available to answer such questions. Knowledge of common signaling pathways used by Qk in flies, worms and fish should be tapped to answer these questions in mammals.

REFERENCES

1. Bohnsack BL, Lai L, Northrop JL et al. Visceral endoderm function is regulated by quaking and required for vascular development. Genesis 2006; 44(2):93-104.
2. Noveroske JK, Lai L, Gaussin V et al. Quaking is essential for blood vessel development. Genesis 2002; 32(3):218-230.
3. Kondo T, Furuta T, Mitsunaga K et al. Genomic organization and expression analysis of the mouse qkI locus. Mamm Genome 1999; 10:662-669.
4. Wu J, Zhou L, Tonissen K et al. The STAR protein quakingI-5 (QKI-5) has a novel nuclear localization signal and shuttles between the nucleus and the cytoplasm. J Biol Chem 1999; 274(41):29202-29210.
5. Wu JI, Reed RB, Grabowski PJ et al. Function of quaking in myelination: regulation of alternative splicing. Proc Natl Acad Sci USA 2002; 99(7):4233-4238.
6. Hardy RJ, Loushin CL, Friedrich VL Jr et al. Glial cell type-specific expression of QKI proteins is altered in quaking viable mutant mice. J Neurosci 1996; 16(24):7941-7949.
7. Zhao L, Tian D, Xia M et al. Rescuing qkV dysmyelination by a single isoform of the selective RNA-binding protein QKI. J Neurosci 2006; 26(44):11278-11286.
8. Ebersole TA, Rho O, Artz K. The proximal end of mouse chromosome 17: New molecular markers identify a deletion associated with quaking viable. Genetics 1992; 131:183-190.
9. Lorenzetti DL, Antalffy B, Vogel H et al. The neurological mutant quaking[viable] is Parkin deficient. Mamm Genome 2004; 15:210-217.
10. Lorenzetti DL, Bishop C, Justice MJ. Deletion of the Parkin co-regulated gene causes male sterility in the quaking[viable] mouse mutant. Proc Natl Acad Sci USA 2004; 101:8402-8407.
11. Cox RD, Hugill A, Shedlovsky A et al. Contrasting effects of ENU-induced embryonic lethal mutations of the quaking gene. Genomics 1999; 57:333-341.
12. Justice MJ, Bode VC. Three ENU-induced alleles of the murine quaking locus are recessive embryonic lethal mutations. Genet Res 1988; 51:95-102.
13. Li Z, Takakura N, Oike Y et al. Defective smooth muscle development in qkI-deficient mice. Development, Growth and Differentiation 2003; 45(5-6):449-462.
14. Chen T, Richard S. Structure-function analysis of qkI: a lethal point mutation in mouse quaking prevents homodimerization. Mol Cell Biol 1998; 18(8):4863-4871.
15. Shedlovsky A, King TR, Dove WF. Saturation germ line mutagenesis of the murine t region including a lethal allele at the quaking locus. Proc Natl Acad Sci USA 1988; 85:180-184.
16. Tam PP, Behringer RR. Mouse gastrulation: the formation of a mammalian body plan. Mech Dev 1997; 68(1-2):3-25.
17. Flamme I, Frolich T, Risau W. Molecular mechanisms of vasculogenesis and embryonic angiogenesis. J Cell Physiol 1997; 173(2):206-210.
18. Noden DM. Embryonic origins and assembly of blood vessels. Am Rev Respir Dis 1989; 140(4):1097-1103.
19. Hopper AF, Heart NH. Foundations of animal development. New York: Oxford University Press; 1985.
20. Lucitti JL, Jones EA, Huang C et al. Vascular remodeling of the mouse yolk sac requires hemodynamic force. Development. 2007 Sep;134(18):3317-26.
21. Ryder SP, Williamson JR. Specificity of the STAR/GSG domain protein Qk1: implications for the regulation of myelination. RNA. 2004 Sep;10(9):1449-58.
22. Saccomanno L, Loushin C, Jan E et al. The STAR protein QKI-6 is a translational repressor. Proc Natl Acad Sci USA 1999; 96(22):12605-12610.
23. Ryder SP, Frater LA, Abramovitz DL et al. RNA target specificity of the STAR/GSG domain post-transcriptional regulatory protein GLD-1. Nat Struct Mol Biol 2004; 11(1):20-28.
24. Jan E, Motzny CK, Graves LE et al. The STAR protein, GLD-1, is a translational regulator of sexual identity in Caenorhabditis elegans. EMBO J 1999; 18(1):258-269.
25. Risua W. Mechanisms of angiogenesis. Nature 1997; 386:671-674.
26. Carmeliet P. Blood vessels and nerves: common signals, pathways and diseases. Nat Rev 2003; 4(9):710-720.
27. Carmeliet P, Collen D. Transgenic mouse models in angiogenesis and cardiovascular disease. J Pathol 2000; 190(3):387-405.
28. Becker S, Wang ZJ, Massey H et al. A role for Indian hedgehog in extraembryonic endoderm differentiation in F9 cells and the early mouse embryo. Dev Biol 1997; 187(2):298-310.
29. Dyer MA, Farrington SM, Mohn D et al. Indian hedgehog activates hematopoiesis and vasculogenesis and can respecify prospective neurectodermal cell fate in the mouse embryo. Development (Cambridge, England) 2001; 128(10):1717-1730.

30. Carmeliet P, Ferreira V, Breier G et al. Abnormal blood vessel development and lethality in embryos lacking a single VEGF allele. Nature 1996; 380(6573):435-439.
31. Ferrara N, Carver-Moore K, Chen H et al. Heterozygous embryonic lethality induced by targeted inactivation of the VEGF gene. Nature 1996; 380:349-442.
32. Leconte I, Fox JC, Baldwin HS et al. Adenoviral-mediated expression of antisense RNA to fibroblast growth factors disrupts murine vascular development. Dev Dyn 1998; 213(4):421-430.
33. Lee SH, Schloss DJ, Swain JL. Maintenance of vascular integrity in the embryo requires signaling through the fibroblast growth factor receptor. The J Biol Chem 2000; 275(43):33679-33687.
34. Maye P, Becker S, Kasameyer E et al. Indian hedgehog signaling in extraembryonic endoderm and ectoderm differentiation in ES embryoid bodies. Mech Dev 2000; 94(1-2):117-132.
35. Shalaby F, Ho J, Stanford WL et al. A requirement for Flk1 in primitive and definitive hematopoiesis and vasculogenesis. Cell 1997; 89(6):981-990.
36. Shalaby F, Rossant J, Yamaguchi TP et al. Failure of blood-island formation and vasculogenesis in Flk-1-deficient mice. Nature 1995; 376(6535):62-66.
37. Shing Y, Folkman J, Sullivan R et al. Heparin affinity: purification of a tumor-derived capillary endothelial cell growth factor. Science (New York, NY 1984; 223(4642):1296-1299.
38. Patan S. Vasculogenesis and angiogenesis as mechanisms of vascular network formation, growth and remodeling. Journal of Neuro-Oncology 2000; 50(1-2):1-15.
39. Baron MH. Embryonic origins of mammalian hematopoiesis. Exp Hematol 2003; 31(12):1160-1169.
40. Winnier G, Blessing M, Labosky PA et al. Bone morphogenetic protein-4 is required for mesoderm formation and patterning in the mouse. Genes Dev 1995; 9(17):2105-2116.
41. Lai L, Bohnsack BL, Niederreither K et al. Retinoic acid regulates endothelial cell proliferation during vasculogenesis. Development (Cambridge, England) 2003; 130(26):6465-6474.
42. Bohnsack BL, Lai L, Dolle P et al. Signaling hierarchy downstream of retinoic acid that independently regulates vascular remodeling and endothelial cell proliferation. Genes Dev 2004; 18(11):1345-1358.
43. Dumont DJ, Gradwohl G, Fong GH et al. Dominant-negative and targeted null mutations in the endothelial receptor tyrosine kinase, tek, reveal a critical role in vasculogenesis of the embryo. Genes Dev 1994; 8(16):1897-1909.
44. Suri C, Jones PF, Patan S et al. Requisite role of angiopoietin-1, a ligand for the tie2 receptor, during embryonic angiogenesis. Cell 1996; 87:1171-1180.
45. Lindahl P, Johansson BR, Leveen P et al. Pericyte loss and microaneurysm formation in PDGF-B-deficient mice. Science (New York, NY 1997; 277(5323):242-245.
46. Hirschi KK, Rohovsky SA, Beck LH et al. Endothelial cells modulate the proliferation of mural cell precursors via platelet-derived growth factor-BB and heterotypic cell contact. Circ Res 1999; 84(3):298-305.
47. Hirschi KK, Rohovsky SA, D'Amore PA. PDGF, TGF-beta and heterotypic cell-cell interactions mediate endothelial cell-induced recruitment of 10T1/2 cells and their differentiation to a smooth muscle fate. J Cell Biol 1998; 141(3):805-814.
48. Hirschi KK, Burt JM, Hirschi KD et al. Gap junction communication mediates transforming growth factor-beta activation and endothelial-induced mural cell differentiation. Circ Res 2003; 93(5):429-437.
49. Limbourg FP, Takeshita K, Radtke F et al. Essential role of endothelial Notch1 in angiogenesis. Circulation 2005; 111(14):1826-1832.
50. Kwon SM, Eguchi M, Wada M et al. Specific Jagged-1 signal from bone marrow microenvironment is required for endothelial progenitor cell development for neovascularization. Circulation 2008; 118(2):157-165.
51. Fischer A, Schumacher N, Maier M et al. The Notch target genes Hey1 and Hey2 are required for embryonic vascular development. Genes Dev 2004; 18(8):901-911.
52. Bohnsack BL, Hirschi KK. Red light, green light: signals that control endothelial cell proliferation during embryonic vascular development. Cell Cycle (Georgetown, Tex 2004; 3(12):1506-1511.
53. Byrd N, Becker S, Maye P et al. Hedgehog is required for murine yolk sac angiogenesis. Development 2002 Jan;129(2):361-72.
54. Dickson MC, Martin JS, Cousins FM et al. Defective haematopoiesis and vasculogenesis in transforming growth factor-beta 1 knock out mice. Development (Cambridge, England) 1995; 121(6):1845-1854.
55. Goumans MJ, Zwijsen A, van Rooijen MA et al. Transforming growth factor-beta signalling in extraembryonic mesoderm is required for yolk sac vasculogenesis in mice. Development (Cambridge, England) 1999; 126(16):3473-3483.
56. Larsson J, Goumans MJ, Sjostrand LJ et al. Abnormal angiogenesis but intact hematopoietic potential in TGF-beta type I receptor-deficient mice. EMBO J 2001; 20(7):1663-1673.
57. Barsi JC, Rajendra R, Wu JI et al. Mind bomb1 is a ubiquitin ligase essential for mouse embryonic development and Notch signaling. Mech Dev 2005; 122(10):1106-1117.
58. Enciso JM, Hirschi KK. Nutrient regulation of tumor and vascular endothelial cell proliferation. Curr Cancer Drug Targets 2007; 7(5):432-437.

59. Ebersole TA, Chen Q, Justice MJ et al. The quaking gene product necessary in embryogenesis and myelination combines features of RNA binding and signal transduction proteins. Nat Genet 1996; 12:260-265.
60. Hardy RJ. QKI expression is regulated during neuron-glial cell fate decisions. J Neurosci Res 1998; 54:46-57.
61. Hardy RJ. The QkI Gene. In: Lazzarini R, ed. Myelin Biology and Disorders. Vol I. Philadelphia: Elsevier Science; 2004:643-659.
62. Zorn AM, Krieg PA. The KH domain protein encoded by quaking functions as a dimer and is essential for notochord development in Xenopus embryos. Genes Dev 1997; 11:2176-2190.
63. Zorn AM, Grow M, Patterson KD et al. Remarkable sequence conservation of transcripts encoding amphibian and mammalian homologues of quaking, a KH domain RNA-binding protein. Gene 1997; 188:199-206.
64. Zaffran S, Astier M, Gratecos D et al. The held out wings (how) Drosophila gene encodes a putative RNA-binding protein involved in the control of muscular and cardiac activity. Development 1997; 124:2087-2098.
65. Tanaka H, Abe K, Kim C. Cloning and expression of the quaking gene in zebrafish embryo. Mech Dev 1997; 69:209-213.
66. Volk T, Israeli D, Nir R et al. Tissue development and RNA control: "HOW" is it coordinated? Trends Genet 2008; 24(2):94-101.
67. Zacchigna S, Ruiz de Almodovar C, Carmeliet P. Similarities between angiogenesis and neural development: what small animal models can tell us. Current Topics in Dev Biol 2008; 80:1-55.
68. Carmeliet P, Tessier-Lavigne M. Common mechanisms of nerve and blood vessel wiring. Nature 2005; 436(7048):193-200.
69. Sondell M, Sundler F, Kanje M. Vascular endothelial growth factor is a neurotrophic factor which stimulates axonal outgrowth through the flk-1 receptor. Eur J Neurosci 2000; 12(12):4243-4254.
70. Shima DT, Mailhos C. Vascular Dev Biol: getting nervous. Curr Opin Genet Dev 2000; 10(5):536-542.
71. Hartenstein V, Mandal L. The blood/vascular system in a phylogenetic perspective. Bioessays 2006; 28(12):1203-1210.
72. Medioni C, Astier M, Zmojdzian M et al. Genetic control of cell morphogenesis during Drosophila melanogaster cardiac tube formation. J Cell Biol 2008; 182(2):249-261.
73. Legg JA, Herbert JM, Clissold P et al. Slits and Roundabouts in cancer, tumour angiogenesis and endothelial cell migration. Angiogenesis 2008; 11(1):13-21.
74. Helenius IT, Beitel GJ. The first "Slit" is the deepest: the secret to a hollow heart. J Cell Biol 2008; 182(2):221-223.
75. Killeen MT, Sybingco SS. Netrin, Slit and Wnt receptors allow axons to choose the axis of migration. Dev Biol 2008; 323(2):143-151.
76. Noveroske JK, Hardy R, Dapper J et al. A new ENU-Induced allele of the mouse quaking gene causes severe central nervous system dysmyelination. Mamm Genome 2005; 16:672-682.

CHAPTER 7

DROSOPHILA STAR PROTEINS
What Can Be Learned from Flies?

Talila Volk*

Abstract   Signal transduction and activation of RNA (STAR) family of RNA binding proteins
are highly conserved through evolution indicating their core role during development,
as well as in adult life. This chapter focuses on two *Drosophila* STAR proteins: Held
Out Wing (HOW), the ortholog of mammalian Quaking (QKI) and Kep1, one of the
four orthologs of mammalian Sam 68. I will emphasize the orthologs similarities in
splicing pattern, functions and mode of actions of the two proteins relying on recent
and earlier findings in the field. I will start with general description of the STAR
proteins in *Drosophila* with an emphasis on their specific expression patterns.

STAR PROTEINS IN *DROSOPHILA*

The sequence of the *Drosophila* genome and its subsequent gene annotation led to
the identification of six STAR family members in the *Drosophila* genome.[1] These may be
divided into three sub groups according to their similarity to the mammalian STAR proteins.
The first group includes true orthologs of the Sam68/SLM1/2 STAR sub-group (also known
as KHDRBS1, KHDRBS2, KHDRBS3) and contains four of these genes (*qkr54B/Sam50*,
qkr58E-1, *qkr58E2* and *qkr58E-3/kep1*). These genes exhibit varied expression patterns;
Sam50 and *qkr58E-1* are maternally contributed and are then highly expressed in males,
whereas *qkr58E2* is strongly expressed at all stages, including in the adult flies. Among
these, the expression pattern of *kep1* is unique, exhibiting peaks at different developmental
stages. It is provided maternally and it is highly expressed during embryogenesis. During
larval stages the gene is shut off and it reappears during metamorphosis. Kep-1 is involved
in apoptosis as well as in oogenesis as will describe in details below.[2] The second group

*Talila Volk—Department of Molecular Genetics, Weizmann Institute of Science, Rehovot 76100, Israel.
Email: talila.volk@weizmann.ac.il

*Post-Transcriptional Regulation by STAR Proteins: Control of RNA Metabolism in Development
and Disease*, edited by Talila Volk and Karen Artzt.

includes a single representative, *how*, which is a true homolog of mammalian *Quaking* (*Qk*) and of *C. elegans asd-2*.[3] *How* is provided maternally, however its zygotic expression in later developmental stages is spatially and temporally regulated; the gene product is detected in a restricted number of tissues, including muscles, tendons and glial cells. The third group includes the *Drosophila* ortholog of the mammalian splicing factor, *SF1*.

The restricted number of STAR family members in the *Drosophila* genome in addition to the availability of tissue specific markers and advanced genetic tools, make it an excellent model organism in which to study the biological significance of the function of individual STAR proteins and their contribution to developmental processes as well as to the adult physiology.

HOW REGULATES DIFFERENTIATION OF DIVERSE TISSUES

Structural Hallmarks and Binding Specificity of HOW

An interesting aspect of both *how* and *quaking* genes is their biogenesis as distinct splice variants due to alternate splicing of exons at their 3′ end. The *how* gene is alternatively spliced into three splice variants named according to their length, HOW(L), HOW(M) and HOW(S), similar to *quaking* (see Fig. 1).[4,5] The HOW(L) isoform, which corresponds to QKI-5, is initially detected at early embryonic stages, whereas the onset of HOW(S) expression, which corresponds to QKI-6 and QKI-7, is in later stages of differentiated tissues. The expression pattern of HOW(M), the most recently identified isoform, awaits further characterization. The highly conserved single maxi-KH-RNA-binding domain shared by the STAR proteins suggests that they all recognize similar RNA target sequences. Indeed, Gld-1, QKI and HOW bind to a core sequence of (U/X)ACUA/CA with some differences: for HOW binding, the first residue can be any of the four nucleotides (X), whereas for Gld-1 and QKI the first nucleotide should be U.[6-8] Additional RNA secondary structural requirements for protein binding to RNA might enhance the binding affinity and/or specificity, presumably due to favorable positioning of the site within the entire 3′UTR. For example, HOW/RNA binding is optimal when the consensus binding site is located on a single stranded loop of at least 12 nt, whereas no binding is detected when the response element is located on a stem, or within a small loop (<12 nt)[8](Fig. 2). Multiple binding sites within the target RNA may also enhance the STAR–RNA binding affinity.

In summary, both the binding specificity and biogenesis of the distinct isoforms are shared between HOW and mammalian QKI.

Functional Analysis of HOW in the *Drosophila* Embryo

Like Quaking, HOW is prominently expressed in a tissue-restricted pattern during embryonic development, as well as in the adult fly.[9-11] In each tissue, it appears to regulate distinct targets; thus, the identification of the set of HOW-targets in different tissues remains a great challenge that should help to elucidate the molecular basis for HOW activity. Analyses performed so far on HOW function during embryonic development revealed distinct paradigms of HOW functions, depending on the tissue and the responsiveness of its target mRNAs.

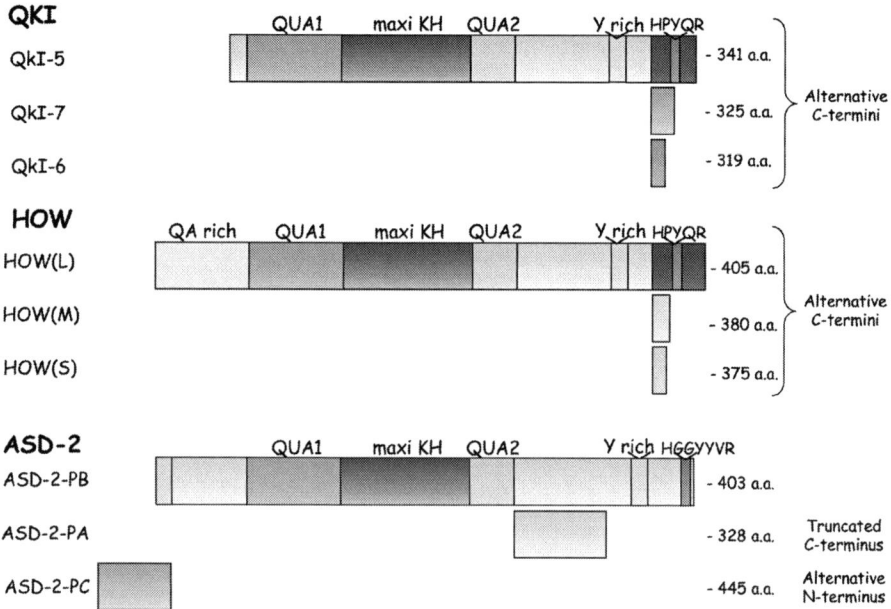

Figure 1. Structural similarities between the STAR proteins QKI, HOW and ASD-2. The maxi KH domain as well as the two QUA1 and QUA2 flanking domains (shown in green) are conserved in the three STAR proteins, mammalian QKI, *Drosophila* HOW and *C. Elegeans* ASD-2 are shown. In addition, the spliced isoforms are indicated. Note that mammalian QKI and *Drosophila* HOW exhibit a similar splicing pattern at the C' terminal domain of the protein, with longer isoforms (QKI-5 and HOW(L)) that share a conserved domain (HPYQR) essential for nuclear retention, a Tyrosine rich domain (Y rich) that might be regulated by phosphorylation by Tyrosine-Kinase and two additional distinct shorter isoforms. In ASD-2 the splicing isoforms and the putative nuclear retention signal are slightly different. A color version of this image is available at www.landesbioscience.com/curie.

Dual Activities of Distinct HOW Isoforms Promote Tendon Cell Differentiation

The activity of HOW during tendon cell differentiation represents an interesting and novel paradigm for regulation of tissue differentiation employed by the interplay between HOW(L) and HOW(S) isoforms. HOW isoforms are identical throughout most of their sequence including the maxi-KH domain, but differ in their C-termini enabling them to promote opposing activities, as will be exemplified bellow.

A critical step during development to achieve active functioning muscles depends on their migration from the mesoderm layer and their attachment to tendons. *Drosophila* tendon cells are specialized ectodermal cells that, upon muscle binding, undergo differentiation into large elongated cells that become filled with polarized arrays of microtubules. At the end of embryogenesis, these cells, as well as neighboring cells, secrete cuticle proteins that together form the exosleleton of the hatching larvae, which, like the skeleton in vertebrates, functions to counteract muscle contraction. A critical stage in the differentiation state of tendon cells is their transition from competent tendon precursors defined in the ectoderm by early patterning genes, into fully differentiated

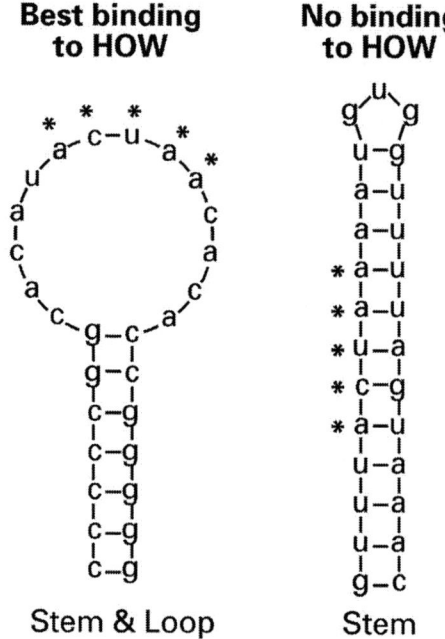

Figure 2. Secondary structure of the optimal binding site for HOW. The consensus RNA HOW-binding sequence (designated by *) is shown as formed by two alternate RNA secondary structure configurations: a loop of 12 nucleotides formed as a result of neighboring stem (left, stem and loop structure) and a stem structure (right). Optimal binding of HOW occurs when the consensus site is within a loop.[8] Reprinted with permission from: Volk T et al. Trends Genet 2008; 24:94-101;[16] ©2008 Elsevier.

muscle-bound cells.[12] HOW proteins appear to regulate this differentiation transition by regulating the levels of the tendon's key transcription factor, Stripe.

Stripe is an EGR-like protein, which regulates the identity of tendon cells. The *stripe* gene encodes two isoforms, *stripe A* and *stripe B*, formed by alternate splicing.[13] Stripe B corresponds to the early isoform that defines the identity of tendon precursors in early developmental stages. Stripe A represents an isoform expressed in tendon cells only following their binding to muscle cells and is essential to promote their terminal differentiation into mature tendons.[14] Thus, the transition between cells expressing Stripe B into cells expressing Stripe A is essential to trigger terminal differentiation of tendon cells.

HOW isoforms are highly expressed in tendon cells throughout their differentiation and regulate *stripe* mRNA levels.[5] HOW binds to the 3'UTR of *stripe* mRNA (common to both *stripe* isoforms) at four repeated HOW Response Elements (HREs) and in addition, associates with intronic sequences specific for *stripe A*.[4,8]

In immature tendon cells, StripeB activates the expression of HOW(L), which in turn binds to *stripeB* 3'UTR and reduces its mRNA levels. This creates a negative feedback loop resulting in low StripeB expression in the tendon precursors, presumably required to maintain the tendon precursors in their immature state (Fig. 3). Muscle binding to tendon cells activates a positive feedback loop supporting StripeA expression. This positive loop is initiated by the activation of the EGF receptor (EgfR) signaling pathway within the tendon cell (as a result of muscle binding), which, among other activities, also elevates the levels

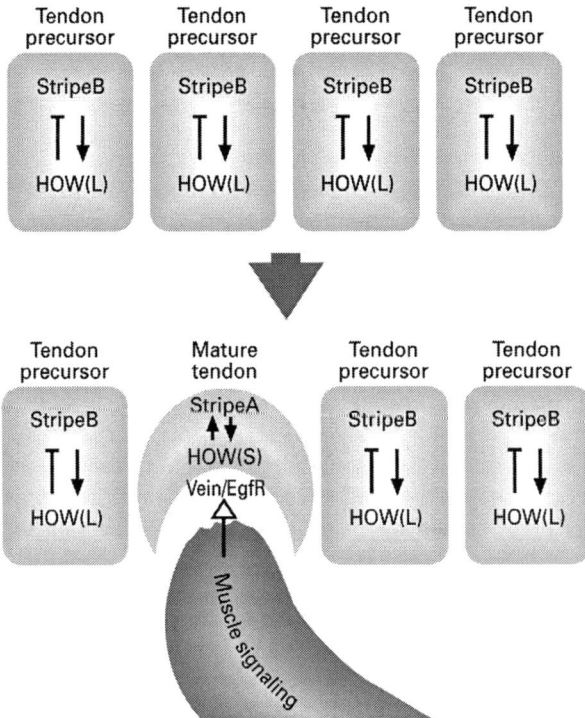

Figure 3. HOW mediates a switch in tendon cell maturation. Tendon cells are specified in the ectoderm by the expression of the transcription factor, StripeB. A negative feedback loop between StripeB and HOW(L) maintains tendons at the precursor stage (upper panel). Somatic muscles migrate towards the tendon precursor cell and provide a differentiation signal, Vein, which binds and activates the EgfR signaling pathway in the tendon cell. This leads to elevation in HOW(S) levels, which then elevate the StripeA isoform characteristic of the mature tendon state. Reprinted with permission from: Volk T et al. Trends Genet 2008; 24:94-101;[16] ©2008 Elsevier.

of the HOW(S) isoform in the muscle-bound tendon cell.[4,15] HOW(S) in turn increases *stripeA* mRNA and protein levels. Extrapolation from a tissue culture experiment in which HOW(S) elevated the levels of a *stripeA* minigene, suggests that HOW(S) elevates the levels of StripeA by promoting its splicing.[14] Therefore, the HOW-induced switch in tendon cell differentiation depends on the relative levels of the repressor isoform, HOW(L) and the promoting isoform, HOW(S) (Fig. 3).

In summary, in tendon cells, HOW proteins act in two distinct regulatory pathways. Prior to muscle binding, HOW(L) is involved in a negative feedback loop, which tempers the instructive effect of StripeB, thereby maintaining the tendon precursor state. Later, upon tendon-muscle attachment, HOW(S) mediates the elevation of *stripeA* mRNA in response to tendon-muscle binding, thereby enabling resumption of tendon cell maturation (Fig. 3).[16] Can we extrapolate from the dual activities of HOW proteins during tendon maturation to the activities of the QKI proteins isoforms during oligodendrocyte differentiation?

Perhaps in precursor oligodendrocytes, the QKI-5 isoform represses differentiation, presumably inhibiting the expression of genes required for terminal differentiation such as *myelin basic protein* (*Mbp*), *myelin protein zero* (*Mp0*) and others, whereas following induction of oligodendrocyte maturation, the levels of QKI-6 and QKI-7 increase, leading to elevation of their target mRNA, essential for terminal differentiation.[17-19] Further functional studies of QKI isoforms are needed to verify this.

Repression of Gene Expression during Mesoderm Development

Another mode of regulation exhibited by HOW is repression of gene expression by inducing destabilization of specific target mRNAs. This activity is performed by maternally contributed as well as zygotic expression of HOW during early mesoderm development. The zygotic transcription of *how(L)* is promoted by Twist, a bHLH transcription factor that specifies the mesoderm germ layer.[20] HOW is essential for early mesoderm development, as embryos lacking both maternal and zygotic *how* exhibit defects in mesoderm spreading, leading to a sporadic lack of dorsal mesoderm derivatives, including heart and dorsal muscles.[21]

To identify mRNA targets whose expression might be repressed by HOW during mesoderm spreading, a comparison between the mRNA profile of *how* mutant embryos versus wild type embryos was performed. Genes that exhibited changes in their mRNA expression levels were then screened for the existence of the HRE in their 3'UTRs and subsequently assayed for the direct binding of HOW to these 3'UTRs.[22] Among the mRNAs identified was *miple1,* whose levels are elevated about eight fold in the mesoderm of *how* mutant embryos. *Miple1* and *Miple2* are the *Drosophila* orthologs of the Midkine/Pleiotrophin family, whose vertebrate homologs activate receptor tyrosine kinases.[23] Repression of *miple1* mRNA is essential for mesoderm spreading and mesodermal Miple1 overexpression therefore phenocopies the defects in mesoderm spreading, as well as the occasional MAPK activation that is also observed in *how* mutants. Additional experiments showed that HOW binds to *miple1* 3'UTR and that its mRNA levels are indeed elevated in *how* mutant embryos.[22] Taken together, these results suggested that HOW directly represses *miple1* expression in the mesoderm of wild-type embryos to enable correct mesoderm spreading. In this model, HOW functions in a temporal manner to negatively regulate a set of genes transcribed by a potent transcription factor. The expression of such genes might interfere with normal development at certain specific developmental stages and thus, expression of these genes must be temporarily repressed.

Facilitation of Tissue Differentiation

Whereas in early developmental stages HOW appears to repress tissue differentiation, in later stages HOW appears to promote tissue differentiation. This has been described in two distinct tissues, glial cells and their ability to wrap and insulate axon bundles and dorsal vessel cardioblasts and their ability to form a lumen.

HOW Function in Glial Cell Maturation and Axon Wrapping. How is highly expressed in several types of glia cells including midline glia, subperineurial glia and perineurial glia.[24] Like QKI in mice, HOW is essential for glial cell maturation. The repertoire of mRNAs regulated by HOW in glial cells has not yet been characterized. In *Drosophila,* glial cells are generated in the central nervous system (CNS) and then migrate along axonal tracts. Once the entire network of neuronal connectivity has been established, the glial cells wrap the axons tightly to ensure their insulation from the high potassium content in the extracellular environment.[25] Whereas this insulation does not involve myelin production, it is functionally analogous to the mammalian myelin sheath and is essential for the efficient and fast conductance of the action potential within axons. The mechanisms directing the arrest of glial cell differentiation prior to and its resumption after the completion of neuronal connectivity are currently unknown.

Recent genetic evidence suggests that maturation of glial cells depends on the activity of HOW together with the splicing factor, Crooked neck (Crn). In both *crn* and *how* null mutants, peripheral glial cells fail to wrap the axons, although these axons have completed their extension and target recognition.[26] A major target for HOW regulatory activity is the *neurexinIV* (*nrxIV*) pre mRNA. NrxIV is an essential component of the septate junctions formed between glial cells and provides axon insulation similar to the blood brain barrier (BBB) in the mammalian nervous system. Mutations in *nrxIV* disrupt both nerve insulation and the formation of the *Drosophila* BBB.[27] The *nrxIV* gene is alternatively spliced into at least two splice variants. Only one of these variants is essential for septate junction formation.[28] In both *how* and *crn* mutant embryos, NrxIV is specifically reduced in septate junctions formed by mature peripheral glia cells, leading to aberrant nerve insulation. Studies in S2 cells showed that Crn and HOW(S) form a protein complex that shuttles to the nucleus and mediates alternative splicing of a *nrxIV* minigene. Moreover, HOW is capable of inducing splicing of a *nrxIV* minigene in S2 cells. Thus, the interaction of HOW with specific partners may contribute to differential regulation of the target RNA.[26]

Based on the *how* and *crn* phenotypes, as well as other experiments, it was suggested that HOW and Crn mediate alternative splicing of the *nrxIV* splice variant characteristic of mature glia. This implicates HOW in controlling proper glial cell maturation; however, the exact function of HOW in the splicing of *nrxIV* as well as the temporal regulation of this event are yet to be elucidated.

Interestingly, QKI was shown to promote oligodendrocyte maturation by inducing the splicing of MAG pre-mRNA.[29] Although there are fundamental differences between *D. melanogaster* glial cells and mammalian oligodendrocytes in their structure and lack of myelin production, both cell types use a homologous STAR protein to control the transition to a glial maturation program. This demonstrates evolutionary conservation of their function in the temporal control of glial cell development.

HOW and Heart Lumen Formation. HOW is highly expressed in cardiac muscles.[5,9] Using a hypomorphic weak allele *how*[18] in which mesoderm development is normal, it was shown that HOW promotes lumen formation in the "dorsal vessel", an organ that includes both the heart and the aorta.[30] This organ consists of a simple tube opened in one side that, upon contraction, pumps out the hemolymph throughout the entire body of the embryo. The lumen of the dorsal vessel is created by special morphogenesis of two rows of cardioblasts that form contacts only at the apical and basal tips while their apical surfaces form the lumen (Fig. 5). HOW is required for the formation of this lumen and appears to regulate the expression/distribution of Slit, shown to be required for lumen formation.[30] It is still not clear whether *slit* RNA is directly regulated by HOW and at which stage this regulation occurs. This might be considered a functional counterpart of the role of QKI in creation and maintenance of a circulatory system in mammals, the failure of which is responsible for *Qk/Qk* embryonic death.

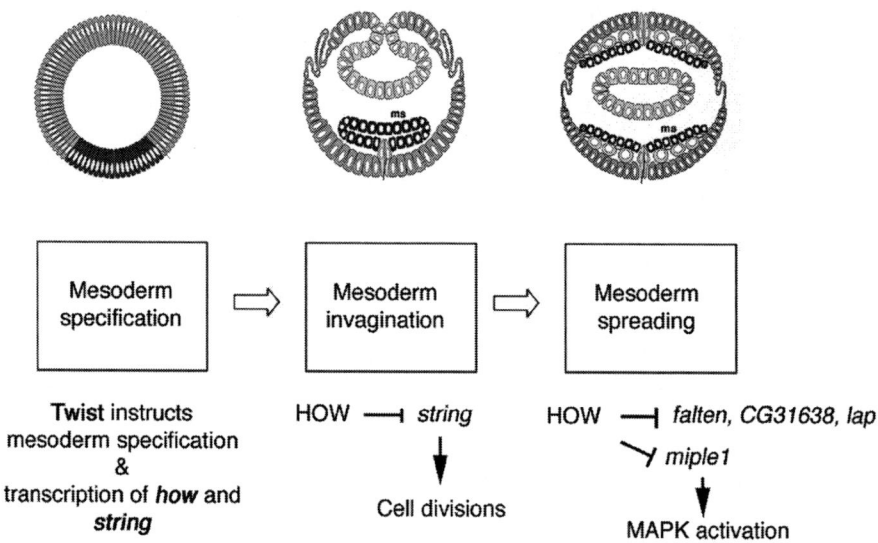

Figure 4. HOW is essential for mesoderm invagination and spreading. Schematic representation of three stages in early mesoderm (black) development: mesoderm specification, mesoderm invagination and mesoderm spreading; the corresponding involvement of HOW at each stage is shown. First, Twist instructs mesoderm specification as well as transcription of *how* and *string*. During mesoderm invagination, HOW represses *string* mRNA levels, arresting cell cycle progression. During mesoderm spreading, HOW represses the mRNA levels of several genes, including *falten*, *lap* and *miple1*. The repression of *miple1* is essential to maintain MAPK activation only in the most dorsal row of mesodermal cells.[22] Reprinted with permission from: Volk T et al. Trends Genet 2008; 24:94-101;[16] ©2008 Elsevier. A color version of this image is available at www.landesbioscience.com/curie.

HOW AND KEP1 REGULATE CELL DIVISION AND APOPTOSIS IN *DROSOPHILA*

HOW Mediates Cell Cycle Arrest in the Mesoderm of Gastrulating Embryos

The involvement of HOW in cell cycle arrest has been demonstrated in gastrulation of the *Drosophila* embryo. During this process, the mesoderm invaginates from the ventral aspects of the blastoderm stage embryo to form a distinct internal germ layer, the mesoderm (Fig. 4). The invagination of the mesoderm involves significant changes in cell shape, during which cells must temporally arrest cell cycle progression.[31] Embryos lacking both maternal and zygotic HOW exhibit multiple ectopic cell divisions in the mesoderm. In these embryos, mesoderm invagination is delayed and is unsynchronized.[21] It appears that the invagination defect in *how* mutants stems from the lack of HOW(L), which is highly expressed at this developmental stage. HOW(L) negatively regulates the mRNA levels of the cell cycle promoting protein, String [also called Cdc25 (Cell division control 25)], during gastrulation. String is a dual specificity phosphatase that positively regulates the cell cycle by activating the mitotic kinase Cdc2 at the G2/M transition.[32,33] In postblastoderm stages, String is the limiting factor that regulates the precise timing and sites of cell division. Once expressed, String protein is constitutively active; therefore, its levels must be tightly regulated. Both

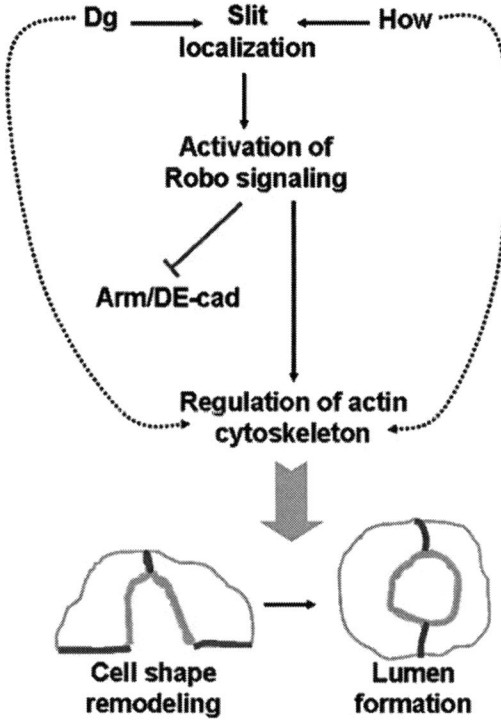

Figure 5. Model for HOW activity during cardiac tube formation. Schematic representation of the interactions between Slit, Robo, HOW and Dg in Cardioblasts that control lumen formation and growth. Dotted arrows, hypothetic interactions. © Medioni et al, 2008. Originally published in *The Journal of Cell Biology*, doi: 10,1083/jcb.200801100.[30] A color version of this image is available at www.landesbioscience.com/curie.

string mRNA and protein are extremely labile (half-life is <15 minutes).[32] The elevation of *string* causes premature cell division during the invagination process, which impairs the synchronization and timing of mesoderm formation.[31] Importantly, *string* contains a consensus HRE in its 3′UTR, shown to be bound directly by HOW.

Twist, the major regulator of mesoderm formation, positively regulates *string* and thus can potentially induce premature cell divisions prior to mesoderm invagination. Twist also induces the expression of *how* in the mesoderm. The parallel Twist-dependent activation of *how*, together with pre-existing HOW from maternal sources, leads to temporal reduction of *string* levels, thereby enabling the mesodermal cells that remain arrested in cell-cycle progression to invaginate from the ectoderm[21] (Fig. 4). HOW, in this case, modulates the instructive activity of Twist by temporal repression of one of its downstream target genes, *string*, delaying cell division in the invaginating mesoderm. Cell-cycle progression resumes once the mesoderm invaginates into the interior of the embryo. Following mesoderm invagination, *string* mRNA accumulates in response to Twist transcriptional activity, thereby enabling sufficient String production to promote cell cycle progression. Thus, the balance between the transcriptional input by Twist and the relative amount of HOW(L) protein levels determines the extent of cell cycle progression during mesoderm development.

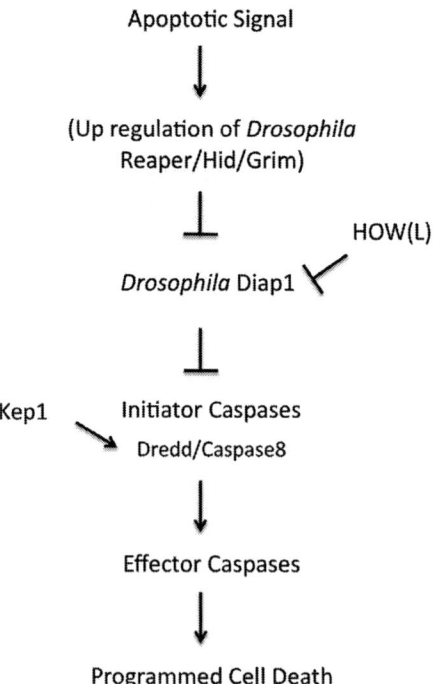

Figure 6. The involvement of HOW and Kep1 in programmed cell death (PCD). Schematic representation of the major players in PCD in *Drosophila*. The function of the two STAR proteins HOW and Kep1 in this regulation is shown. HOW reduces the levels of *Diap1* mRNA and Kep1 induces the correct splicing of *Dredd/Cas8*.

Kep1 and HOW Promote the Apoptotic Process in *Drosophila*

Apoptosis in *Drosophila* is primarily mediated by the three apoptotic gene products, Reaper, Hid and Grim, which initiate apoptosis by inhibiting the activity of the Inhibitors of Apoptosis (IAPs).[34,35] *Drosophila* inhibitor of apoptosis (Diap1) inhibits both the initiator and effector Caspases[36] (Fig. 6). QKI as well as Sam68 were shown to be involved in apoptosis in mammalian cells in culture, however the contribution of these genes to cell death in the whole organism remains elusive.[37-39] In *Drosophila*, both Kep1 and HOW were shown to be positive regulators of apoptosis; however, their target mRNAs differ. Kep1 appears to promote the alternative splicing of an active isoform of the caspase-8-like *dredd*.[2] HOW appears to regulate the mRNA levels of *diap1*.[24] In addition, Kep1 is active during oogenesis, whereas HOW promotes apoptosis in midline glial cells. *kep−/−* flies are viable but display reduced fertility in females. The un-hatched eggs display variable phenotypes including shape changes and lack of appendages. The mutant ovaries display reduced activity of Dredd/Caspase-8 and the mRNA levels of the active *dredd* isoforms are down regulated. Significantly, Kep1 binds to *dredd* pre-mRNA and its deficiency prevents *rpr*-induced apoptosis in the eye of the adult fly.[2] These data support the notion that Kep1 positively regulates apoptosis by promoting the splicing of the active isoform of Dredd/Caspase-8. Kep1 may not be constitutively active in the ovary and its tyrosine phsophorylation by Src appears to facilitate the association of Kep1 with the splicing factor, SF2, promoting efficient splicing.

HOW was shown to mediate apoptosis in the midline glia (MG) cells of the *Drosophila* embryo. These cells are essential to promote axonal pathfinding across the midline and to separate between the anterior and posterior commissures. In early developmental stages, six MG cells are present. These cells undergo selective cell death during their migration to the midline resulting in three surviving cells at Stage 16.[40,41] In *how* mutant embryos, the number of MG cells remains high, leading to an aberrant pattern of the ventral cord commissures. This phenotype can be rescued by midline expression of the HOW(L) isoform. The basis for the HOW-dependent apoptosis of MG cells appears to be its negative regulation of the mRNA levels of the caspase inhibitor of apoptosis, *diap1*. In *how* mutant embryos, the levels of Diap1 are elevated, in parallel to reduction in the levels of the activated effector caspase 3. Accordingly, reducing the levels of HOW in S2 cells leads to elevation of Diap1, whereas over expression of HOW(L) promotes reduction of Diap1 protein as well as mRNA levels. Importantly, deletion of the two HREs from *diap1* 3′UTR abrogates HOW-dependent repression of Diap1, suggesting that HOW represses *diap1* by direct binding to its 3′UTR. Therefore, HOW(L) enhances the sensitivity of MG cells to apoptotic signals by reducing the levels of *diap1* in these cells.[24]

In summary, HOW and Kep1 appear to promote apoptosis by regulating different RNA targets. Kep1 positively regulates the splicing of an active isoform of dredd/ caspase-8, whereas HOW negatively regulates the levels of the inhibitor of apoptosis, Diap1. Importantly, the apoptosis-dependent regulation of each of these proteins depends primarily on the initial activity of the pro-apoptotic genes *reaper*, *hid* and *grim*; thus, their activity is exhibited mainly in the modulation of a pre existing apoptotic signal.

CONCLUSION

Among other RNA binding proteins, STAR proteins affect protein synthesis and mRNA stability rather than controlling the transcription of target genes, enabling a potentially rapid regulatory mechanism of gene expression. In addition, this mechanism allows differential gene regulation in diverse tissues while presumably keeping the expression driven by transcription factors constant. The restricted number of *Drosophila* STAR proteins and their distinct tissue distribution suggest that they display nonredundant functions during embryonic and adult life. The wide range of phenotypes observed in *how* or *kep 1* mutant flies suggests the functional importance of these gene products in the distinct tissues.

Several open questions arise regarding the function of STAR proteins: First, what mechanism regulates their activity in distinct tissues. One level of regulation was shown in tendon cells maturation where signaling mediated by the EGF receptor pathway led to differential expression of each HOW isoform. In addition, several phosphorylation sites for different kinases (MAPK and Src kinase) are present in the coding sequence of both HOW and Kep 1. It will be of great importance to define whether these sites are indeed phosphorylated in vivo and whether this phosphorylation affects their activity. Importantly, this information might reveal the link between various signal transduction pathways and STAR proteins in vivo. An additional open question is the contribution of the different spliced forms to distinct developmental processes as well as the mechanism regulating their differential expression.

In summary, the study of STAR proteins in *Drosophila* has provided novel insights into their function during the development and differentiation of distinct tissues. Future studies are expected to define which of these functions are conserved in higher organisms.

NOTE ADDED IN PROOF

A recent study of HOW function in fly spermatogenesis indicated that HOW is essential to specify the four spermatogonial transit-amplifying divisions in the testis by regulating both CyclinB as well as Bag of marbels (Bam) mRNA levels. This suggests a novel and important function for HOW in stem cell maintenance in the male germline.[42]

ACKNOWLEDGEMENT

I thank H. Toledano-Katchalsky and S. Schwarzbaum for comments on the manuscript. This work is supported by a grant from the German-Israeli Fund (GIF) (I-921-159.3/2006) and a grant from MINERVA STIFTUNG.

REFERENCES

1. Gamberi C, Johnstone O, Lasko P. Drosophila RNA binding proteins. Int Rev Cytol 2006; 248:43-139.
2. Di Fruscio M et al. Kep1 interacts genetically with dredd/caspase-8 and kep1 mutants alter the balance of dredd isoforms. Proc Natl Acad Sci USA 2003; 100:1814-9.
3. Ohno G, Hagiwara M, Kuroyanagi H. STAR family RNA-binding protein ASD-2 regulates developmental switching of mutually exclusive alternative splicing in vivo. Genes Dev 2008; 22:360-74.
4. Nabel-Rosen H, Volohonsky G, Reuveny A et al. Two isoforms of the Drosophila RNA binding protein, how, act in opposing directions to regulate tendon cell differentiation. Dev Cell 2002; 2:183-93.
5. Nabel-Rosen H, Dorevitch N, Reuveny A et al. The balance between two isoforms of the Drosophila RNA-binding protein how controls tendon cell differentiation. Mol Cell 1999; 4:573-84.
6. Galarneau A, Richard S. Target RNA motif and target mRNAs of the Quaking STAR protein. Nat Struct Mol Biol 2005; 12:691-8.
7. Ryder SP, Frater LA, Abramovitz DL et al. RNA target specificity of the STAR/GSG domain post-transcriptional regulatory protein GLD-1. Nat Struct Mol Biol 2004; 11:20-8.
8. Israeli D, Nir R, Volk T. Dissection of the target specificity of the RNA-binding protein HOW reveals dpp mRNA as a novel HOW target. Development 2007; 134:2107-14.
9. Zaffran S, Astier M, Gratecos D et al. The held out wings (how) Drosophila gene encodes a putative RNA-binding protein involved in the control of muscular and cardiac activity. Development 1997; 124:2087-98.
10. Lo PC, Frasch M. A novel KH-domain protein mediates cell adhesion processes in Drosophila. Dev Biol 1997; 190:241-56.
11. Baehrecke EH. who encodes a KH RNA binding protein that functions in muscle development. Development 1997; 124:1323-32.
12. Volk T. Singling out Drosophila tendon cells: a dialogue between two distinct cell types. Trends Genet 1999; 15:448-53.
13. Frommer G, Vorbruggen G, Pasca G et al. Epidermal egr-like zinc finger protein of Drosophila participates in myotube guidance. EMBO J 1996; 15:1642-9.
14. Volohonsky G, Edenfeld G, Klambt C et al. Muscle-dependent maturation of tendon cells is induced by post-transcriptional regulation of stripeA. Development 2007; 134:347-56.
15. Yarnitzky T, Min L, Volk T. The Drosophila neuregulin homolog Vein mediates inductive interactions between myotubes and their epidermal attachment cells. Genes Dev 1997; 11:2691-700.
16. Volk T, Israeli D, Nir R et al. Tissue development and RNA control: "HOW" is it coordinated? Trends Genet 2008; 24:94-101.

17. Cox RD et al. Contrasting effects of ENU induced embryonic lethal mutations of the quaking gene. Genomics 1999; 57:333-41.
18. Chen Y, Tian D, Ku L et al. The selective RNA-binding protein quaking I (QKI) is necessary and sufficient for promoting oligodendroglia differentiation. J Biol Chem 2007; 282:23553-60.
19. Hardy RJ. Molecular defects in the dysmyelinating mutant quaking. J Neurosci Res 1998; 51:417-22.
20. Furlong EE, Andersen EC, Null B et al. Patterns of gene expression during Drosophila mesoderm development. Science 2001; 293:1629-33.
21. Nabel-Rosen H, Toledano-Katchalski H, Volohonsky G et al. Cell divisions in the drosophila embryonic mesoderm are repressed via post-transcriptional regulation of string/cdc25 by HOW. Curr Biol 2005; 15:295-302.
22. Toledano-Katchalski H, Nir R, Volohonsky G et al. Post-transcriptional repression of the Drosophila midkine and pleiotrophin homolog miple by HOW is essential for correct mesoderm spreading. Development 2007; 134:3473-81.
23. Englund C, Birve A, Falileeva L et al. Miple1 and miple2 encode a family of MK/PTN homologues in Drosophila melanogaster. Dev Genes Evol 2006; 216:10-8.
24. Reuveny A, Elhanany H, Volk T. Enhanced sensitivity of midline glial cells to apoptosis is achieved by HOW(L)-dependent repression of Diap1. Mech Dev 2009; 126:30-41.
25. Stork T et al. Organization and function of the blood-brain barrier in Drosophila. J Neurosci 2008; 28:587-97.
26. Edenfeld G et al. The splicing factor crooked neck associates with the RNA-binding protein HOW to control glial cell maturation in Drosophila. Neuron 2006; 52:969-80.
27. Bhat MA et al. Axon-glia interactions and the domain organization of myelinated axons requires neurexin IV/Caspr/Paranodin. Neuron 2001; 30:369-83.
28. Stork T et al. Drosophila Neurexin IV stabilizes neuron-glia interactions at the CNS midline by binding to Wrapper. Development 2009; 136:1251-61.
29. Wu JI, Reed RB, Grabowski PJ et al. Function of quaking in myelination: regulation of alternative splicing. Proc Natl Acad Sci USA 2002; 99:4233-8.
30. Medioni C, Astier M, Zmojdzian M et al. Genetic control of cell morphogenesis during Drosophila melanogaster cardiac tube formation. J Cell Biol 2008; 182:249-61.
31. Grosshans J, Wieschaus E. A genetic link between morphogenesis and cell division during formation of the ventral furrow in Drosophila. Cell 2000; 101:523-31.
32. Edgar BA, Lehman DA, O'Farrell PH. Transcriptional regulation of string (cdc25): a link between developmental programming and the cell cycle. Development 1994; 120:3131-43.
33. Lehner CF. Pulling the string: cell cycle regulation during Drosophila development. Semin Cell Biol 1991; 2:223-31.
34. Xu D et al. Genetic control of programmed cell death (apoptosis) in Drosophila. Fly (Austin) 2009; 3:78-90.
35. Steller H. Regulation of apoptosis in Drosophila. Cell Death Differ 2008; 15:1132-8.
36. Dotto GP, Silke J. More than cell death: caspases and caspase inhibitors on the move. Dev Cell 2004; 7:2-3.
37. Pilotte J, Larocque D, Richard S. Nuclear translocation controlled by alternatively spliced isoforms inactivates the QUAKING apoptotic inducer. Genes Dev 2001; 15:845-58.
38. Taylor SJ, Resnick RJ, Shalloway D. Sam68 exerts separable effects on cell cycle progression and apoptosis. BMC Cell Biol 5, 5 (2004).
39. Polotskaia A et al. Regulation of arginine methylation in endothelial cells: role in premature senescence and apoptosis. Cell Cycle 2007; 6:2524-30.
40. Granderath S, Klambt C. Glia development in the embryonic CNS of Drosophila. Curr Opin Neurobiol 1999; 9:531-6.
41. Jacobs JR. The midline glia of Drosophila: a molecular genetic model for the developmental functions of glia. Prog Neurobiol 2000; 62:475-508.
42. Monk AC, Siddall NA, Volk T et al. HOW is required for stem cell maintenance in the Drosophila testis and for the onset of transit-amplifying divisions. Cell Stem Cell 2010; 6:348-360.

CHAPTER 8

C. ELEGANS STAR PROTEINS, GLD-1 AND ASD-2, REGULATE SPECIFIC RNA TARGETS TO CONTROL DEVELOPMENT

Min-Ho Lee* and Tim Schedl

Abstract: A comprehensive understanding of the *C. elegans* STAR proteins GLD-1 and ASD-2 is emerging from a combination of studies. Those employing genetic analysis reveal in vivo function, others involving biochemical approaches pursue the identification of mRNA targets through which these proteins act. Lastly, mechanistic studies provide the molecular pathway of target mRNA regulation.

MULTIPLE FUNCTIONS OF GLD-1 IN GERMLINE DEVELOPMENT

The *gld-1* gene (GermLine development Defective) was identified by forward genetic screens to isolate mutations resulting in sterile hermaphrodites or altered hermaphrodite germline sex determination.[1] Multiple functions of *gld-1* were revealed through various mutant phenotypes that disrupted different aspects of germline development. In *C. elegans* there are two sexes, self-fertile hermaphrodites and males.[2] Hermaphrodites are somatically female with a germline that first undergoes male development (spermatogenesis) during late larval stages and then undergoes female development (oogenesis) in adulthood.[3] The wild-type adult hermaphrodite germline resides within two U-shaped gonads with the different germ cell types spatially arrayed in a distal to proximal developmental gradient. For a single gonad (Fig. 1), at the distal end is a population of mitotically dividing cells, including germline stem cells, which enter meiotic prophase in the transition zone (leptotene/zygotene). As these germ cells move proximally, they progress through an extended pachytene region, followed by diplotene and then diakinesis, where morphological differentiation of oocytes occurs. In a sequential fashion the most proximal

*Corresponding Author: Min-Ho Lee—Department of Biological Sciences, University at Albany, SUNY, Albany, New York 12222 Email: mhlee@albany.edu

Post-Transcriptional Regulation by STAR Proteins: Control of RNA Metabolism in Development and Disease, edited by Talila Volk and Karen Artzt.

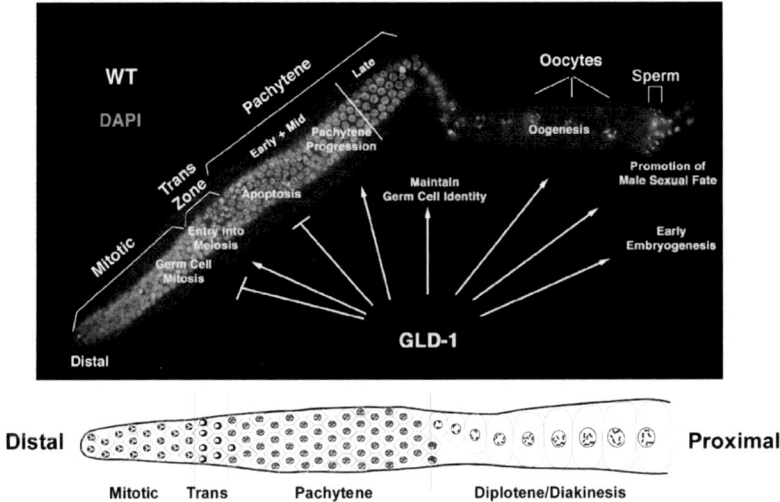

Figure 1. *gld-1* has various functions in *C. elegans* germline development. A dissected germline from a wild-type adult hermaphrodite is stained with DAPI to visualize DNA (light grey or blue, upper panel). Composite shows a surface view of the distal region (left) and an interior focal plane in the proximal region (right). The germline functions of GLD-1 are indicated on the various positions where each developmental decision likely occurs. Diagram of a single wild-type adult hermaphrodite gonad arm (lower panel) is drawn linearly instead of its normal reflex shape for comparison purpose. The adult hermaphrodite germline contains about 1000 germ nuclei. In the distal region, nuclei are arranged primarily around the periphery of the gonadal tube. Each nucleus is partially enclosed by plasma membranes; although this is a syncytium, each nucleus and its surrounding cytoplasm and membranes is called a germ cell. See text for details. A color version of this image is available at www.landesbioscience.com/curie.

oocyte undergoes meiotic maturation, is ovulated, then fertilized by sperm that reside in the spermatheca and begins embryonic development.

Among multiple developmental processes underlying meiotic prophase progression and oogenesis, those particularly relevant to *gld-1* function include: (A) mRNA synthesis primarily occurring in pachytene to be supplied for oocyte development and/or for embryogenesis as maternal mRNAs.[4,5] (B) Apoptosis occurs in substantial portion of germ cells in the proximal half of the pachytene region, as they appear to function as nurse cells.[6] Additional apoptosis can occur in this region if there is unrepaired DNA damage, either generated by irradiation or because of failure to complete meiotic recombination.[7]

gld-1 has an essential function in promoting meiotic prophase progression during oogenesis (Fig. 1).[1,8,9] In *gld-1* null hermaphrodites, the distal most germ cells proliferate and enter meiotic prophase normally; however, germ cells in pachytene exit meiotic prophase, proliferate ectopically leading to the formation of a tumor in the proximal part of the germline. They consequently fail to undergo oocyte development (Fig. 2C). In contrast, germline development in *gld-1* null males is normal, indicating *gld-1* has no essential function in male germline. Some partial loss-of-function alleles show pachytene arrest, while others form abnormal oocytes, consistent with a role for *gld-1* in meiotic prophase progression and oocyte development.[1] Subsequent studies by Ciosk et al[10] found that some cells in the pachytene region, where return to proliferation is occurring, express markers and display morphology consistent with muscle and neuronal cell development, as well as loss of expression of germ cell specific P-granule components. Additionally, it

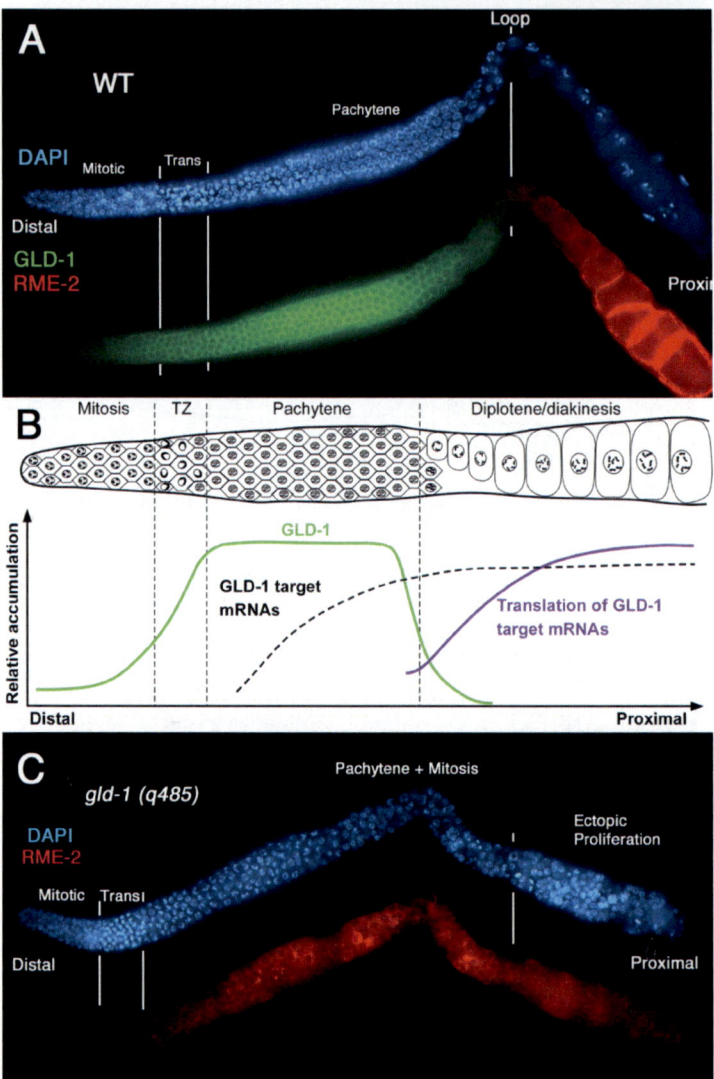

Figure 2. GLD-1 acts as a translational repressor. RME-2 yolk receptor accumulation is regulated by GLD-1. Gonad arms, dissected from (A) a wild-type adult hermaphrodite or from (C) a *gld-1* null adult hermaphrodite, are stained with DAPI (blue) to visualize DNA, anti-GLD-1 antibody (green) and anti-RME-2 antibody (red). (B) Schematic representation of the adult hermaphrodite germline and qualitative depiction of GLD-1 protein (green solid line) levels and mRNA (black dashed line) and protein (purple solid line) levels of some GLD-1 mRNA targets such as RME-2, OMA-1/-2 and PUF-5. Our qualitative assessment of mRNA and protein levels (y-axis) in the corresponding regions of the germline (x-axis) is shown in the graph. GLD-1 and RME-2 accumulation are mutually exclusive in wild-type germlines. Composite shows a surface view of the distal region (left) and an interior focal plane in the proximal region (right). GLD-1 staining is the strongest in the transition zone and pachytene region. At the loop, as GLD-1 staining decreases rapidly, RME-2 staining starts to appear. RME-2 levels increase and are localized at the plasma membrane as proximal oocytes become more fully cellularized and increase in volume in the proximal region. RME-2 is mis-expressed in early pachytene stage germ cells in the distal region of *gld-1* null adult hermaphrodite germlines. At the proximal end, as germ cells proliferate ectopically, RME-2 staining is variable.

was found that mRNAs corresponding to the earliest zygotic transcripts of the embryo are expressed in the region where return to proliferation occurs.[11] These results indicate that oogenic germ cells can transdifferentiate to somatic cell types in the absence of GLD-1. Thus *gld-1* acts to maintain germ cell identity during oogenesis and *gld-1* null germline tumors may represent a *C. elegans* analog of the mammalian teratoma.[10]

During late pachytene stage of female germ cell development, *gld-1* also functions to inhibit germ cell apoptosis.[12] A temperature sensitive allele (*op236*) of *gld-1* was identified, which has increased apoptosis following irradiation at the permissive temperature (20°C), but, in the absence of irradiation resembles wild type. At the restrictive temperature (25°C), *gld-1(op236)* hermaphrodites display a high level of apoptosis even in the absence of irradiation. Increased apoptosis is not observed in *gld-1* null, possibly because germ cells return to mitosis in early pachytene, prior to the time when they become capable of undergoing apoptosis.

Genetic analysis revealed that *gld-1* functions redundantly with a pair of genes, *gld-2* and *gld-3*, to inhibit the germ cell proliferative fate in the distal mitotic region and/or to promote entry into meiotic prophase.[8,13,14] In double mutants of *gld-1* null with either *gld-2* or *gld-3* null, a germline tumor forms in both hermaphrodites and males and in this tumor, germ cells fail to initiate meiotic development.[8,13-15] The etiology of this tumor formation is thus distinct from *gld-1* null single mutant hermaphrodites where germ cells leave pachytene of oogenesis and return to mitotic proliferation. GLP-1 Notch receptor signaling acts as a major switch controlling the decision between germline stem cell proliferation versus entry into meiotic prophase.[16] Genetic epistasis analysis demonstrates that the redundant *gld-1* and *gld-2/gld-3* pathways function downstream of and are inhibited by GLP-1 Notch signaling.[17,18]

gld-1 also functions to specify the male fate in the hermaphrodite germline.[1,8] This function is most clearly revealed by a haplo-insufficient feminization of the hermaphrodite germline by *gld-1* null alleles and multi-locus deletions. However, *gld-1* action in sex determination is complex. First, while *gld-1* null tumorous hermaphrodites usually do not undergo male germline development, the addition of some morphological markers (e.g., Uncs) that normally do not affect sex determination significantly increases the frequency of male germline development in *gld-1* null hermaphrodite germlines. These morphological markers can slow the pace of larval germ cell development, which may, directly or indirectly, affect the decision to begin male germ cell development. Second, dominant antimorphic poisoning alleles of *gld-1* have been identified that either feminize or masculinize the germline. Third, a few of the feminizing antimorphic poisoning alleles (e.g., *q126*) also feminize the germline of males, even though null alleles do not effect male germline development.

GLD-1 MOLECULAR ANALYSIS

Concrete hypotheses on how *gld-1* governs in *C. elegans* germline development became possible following its molecular cloning demonstrating that it encodes an RNA binding protein that contains a STAR domain and from finding a nonhomogenous *gld-1* protein (GLD-1) distribution in the hermaphrodite germline.[9] First, among the many STAR domain containing proteins across phyla, GLD-1 is one of two proteins most closely related to human/mouse Quaking (QKI) and *Drosophila* held out wings (HOW).[19,20-24] The other protein is ASD-2 (see below). Furthermore, based on a structure of another STAR protein, in complex with RNA splicing factor 1 (SF1, also known as the splicing branch point sequence), the amino acid residues predicted to contact RNA in GLD-1, Quaking and HOW are 84% identical (16 out of 19 residues). In contrast, other STAR domain containing proteins such as the Sam68 subfamily have significant amino acid differences,[25] suggesting that the RNA binding specificity of GLD-1, Quaking and HOW are likely similar to each other. The STAR domain of GLD-1 has been shown to be essential in vivo because all 10 missense mutations identified thus far, which reduce, eliminate or alter GLD-1 function, are in conserved amino acids within the STAR domain.[1,12,19]

Second, GLD-1 is localized in the cytoplasm and its distribution throughout the germline is nonhomogenous.[9] In the wild-type adult hermaphrodite germline, the amount of GLD-1 is low at the distal end and increases to its maximum level in the transition zone where germ cells enter meiotic prophase. GLD-1 level remains high in the cytoplasm of pachytene stage germ cells in the distal region while in proximal pachytene, GLD-1 levels begin to decrease and then fall rapidly in diplotene, as oocytes begin morphological differentiation. They are then undetectable in diakinesis stage oocytes (Fig. 2A,B).

The multiple *gld-1* functions in germline development indicated by the various mutant phenotypes suggest that GLD-1 has multiple mRNA targets, where GLD-1 regulation of individual mRNA targets contribute to one or more of its functions. Additionally, the nonhomogenous distribution of GLD-1, its role as a cytoplasmic RNA binding protein and its oogenesis specific tumor production when *gld-1* function is eliminated, led to a model of how GLD-1 likely functions as an RNA binding protein.[19] The oogenic germline produces many maternal mRNAs during early prophase (primarily in pachytene) that are translationally repressed to prevent premature function. These maternal mRNAs are subsequently translated so that the protein products can be used in late stage oocytes, meiotic maturation, meiotic divisions and/or embryogenesis. Thus GLD-1 would bind and repress the translation of its mRNA targets during early meiotic prophase while translational repression of its targets would be relieved after GLD-1 is eliminated as oocytes grow and differentiate in the proximal region. This model also predicts that the germline tumor, arising from germ cells exiting meiotic prophase, likely occurs as a result of premature translation of GLD-1 mRNA targets. The inappropriate activity of certain prematurely translated GLD-1 targets would, in some way, be incompatible with early meiotic prophase, causing germ cells to leave meiotic prophase and begin ectopic proliferation. Thus for GLD-1, as is true for any RNA binding protein, significant advancement in understanding function and regulation comes from identification and study of their mRNA targets.

mRNA TARGETS: GLD-1 IS A TRANSLATIONAL REPRESSOR

A number of in vivo GLD-1 mRNA targets were identified that co-immunoprecipitate (IP) with functional GLD-1 from wild-type adult hermaphrodite cytosol extracts (see Table 1).[26,27] These 16 mRNA targets are preferentially expressed in the germline and many exhibit essential functions in meiotic prophase progression/oogenesis and early embryogenesis, as would be expected of GLD-1 mRNA targets. Other GLD-1 targets were identified through a yeast three hybrid screen (*tra-2* mRNA),[28] or by candidate gene approaches based on the mutually exclusive expression pattern with GLD-1 and the corresponding mRNA target proteins and/or genetic relationships between *gld-1* and its target genes (*mes-3, glp-1, pal-1, cep-1* and *cye-1* mRNAs, Table 1).[11,12,29-31] Recently, a more systematic approach to identify most GLD-1 mRNA targets was performed

Table 1. GLD-1 mRNA targets

mRNA Targets	Homology	Developmental Function	Expression Mitotic	Expression Early, Mid Pachytene	Expression Late Pachytene	Expression Oocytes	References
rme-2	Yolk receptor	Yolk import into Developing Oocytes	No	No	No	Yes	26,57
tra-2	Transmembrane receptor	Female Cell Fates in Germline and Soma	N/D	N/D	N/D	N/D	28,48
mes-3	Novel protein	Maternally supplied Regulator of Germline Development	Yes	No	Yes	Yes	29
pal-1	Homeodomain, Caudal ortholog	Cell Specification in Early Embryo Development	No	No	No	Limited[1]	31
glp-1	Notch Homolog	Specification of the Germline Proliferative Fate, as well as Cell Fates in the Embryo	Yes	No	No	No	30,33
cep-1	p53 ortholog	Promotes DNA Damage-Induced Apoptosis	Yes	No	Yes	Yes	12,58
oma-1/-2 (moe-1/-2)	Zn-finger (C-x8-C-x5-C-x3-H type)	*oma-1* and *oma-2* Function Redundantly in Oocyte Maturation	No	No	No	Yes	27,34,59
gna-2	N-acetyltransferase	Meiotic division and Early Embryonic Development	No	No	No	Limited[2]	26,27,60

continued on next page

Table 1. Continued

mRNA Targets	Homology	Developmental Function	Expression Mitotic	Expression Early, Mid Pachytene	Expression Late Pachytene	Expression Oocytes	References
puf-5/-6/-7/-10	PUF	*puf-5* and *puf-6/-7/-10* Function Redundantly during Oogenesis	No[3]	No[3]	Yes[3]	Yes[3]	26,33
cye-1	Cyclin E homolog	Necessary for the G1/S-phase Transition and Centrosome Duplication	Yes	No	Yes	Yes	11
cej-1 (*cpg-1*)	Chitin-binding Domain	Early embryo development, functionally redundant with *cpg-2*	N/D	N/D	N/D	N/D	26,61
cpg-2	Chitin-binding Domain	Early embryo development functionally redundant with *cpg-1*	N/D	N/D	N/D	N/D	26,61
lin-45	Raf kinase	Progression through Pachytene, Male Germ Cell Fate	N/D	N/D	N/D	N/D	26,37

using GLD-1 IP followed by a microarray detection strategy. From this study almost all previously known targets were identified but in addition, more than 100 additional targets were discovered (Lee et al, unpublished data).

The identification of multiple GLD-1 mRNA targets made it possible to address several important questions. Does GLD-1 regulate the expression of all targets at the same level of RNA metabolism? As a cytoplasmic RNA binding protein, GLD-1 might regulate the stability and/or translation of its targets after mRNA binding. For example, GLD-1 may repress the translation of all targets, as proposed. Alternatively, GLD-1 may repress the translation of some targets while, for other targets, GLD-1 may activate translation or regulate mRNA stability. For all targets examined in detail thus far, GLD-1 functions as a translational repressor: *tra-2*, *rme-2*, *gna-2*, *oma-1*, *oma-2*, *mes-3*, *pal-1*, *glp-1*, *cep-1* and *cye-1* mRNAs (see Table 1).[11,12,26-31] For most targets the mRNAs are present throughout the germline, whether GLD-1 is present or not. However, their proteins are absent in early meiotic (leptotene/zygotene/pachytene) germ cells of wild-type hermaphrodites where GLD-1 is abundant in the cytoplasm; and are then expressed in the developing diplotene/diakinesis oocytes in the proximal region where GLD-1 is absent (Fig. 2A, B). As expected, in *gld-1* null germlines, the target proteins are mis-expressed in early meiotic germ cells (Fig. 2C).[11,12,26,27,29-31]

Two additional levels of regulation can affect the expression of GLD-1 mRNA targets. First, at the distal end where germ cells are proliferating mitotically, GLD-1 levels are low. Thus GLD-1 mRNA targets are likely translated, which occurs for some targets such as MES-3, GLP-1, CEP-1 and CYE-1. However, other targets such as RME-2, OMA-1/-2 and PUF-5 are not expressed (see Table 1). This difference, however, appears to be independent of GLD-1 but dependent on whether the transcript of each target is present or not.[11,12,26,27,29,30] Second, in proximal diplotene/diakinesis oocytes where GLD-1 is nearly absent, proteins corresponding to most mRNA targets accumulate as their mRNAs are relieved from the GLD-1 dependent translational repression. However there are two exceptions, PAL-1 and GLP-1. As described, PAL-1 and GLP-1 proteins are not expressed in pachytene stage germ cells in the distal region. However, in the proximal region, even though GLD-1 is absent, *pal-1* and *glp-1* mRNA translation continues to be repressed (Table 1). Interestingly, this repression occurs via other RNA binding proteins, MEX-3 for *pal-1* mRNA and PUF-5/-6/-7/-10 for *glp-1* mRNA.[31-33] Thus while the translation of *pal-1* and *glp-1* mRNAs in the distal meiotic region are repressed by GLD-1, these mRNAs are continuously repressed in the proximal region by other RNA binding proteins even though GLD-1 is absent. Importantly, PUF-5 expression (PUF-6/-7/-10 have not yet been examined) is restricted to the proximal germline and this restriction likely occurs through translational repression by GLD-1 since *puf-5* and *puf-6/-7/-10* mRNAs were identified as GLD-1 targets.[26] Therefore, translational repression of *pal-1* and *glp-1* mRNAs requires two distinct regulatory systems that are spatially and temporally separated in the germ line.

Interestingly, among the GLD-1 mRNA targets characterized thus far, subsets of three gene families exist; 2 genes of a 6 member family that contains a "chitin binding" domain, 4 genes of a 10 member *pumilio/fbf* (*puf*) family and 2 closely related zinc finger proteins (*oma-1/-2*)(see Table 1). RNAi studies indicate that the two chitin binding domain containing proteins function redundantly in early embryogenesis, *puf-5/-6/-7/-10* in late oogenesis and *oma-1/-2* in oocyte maturation.[26,27,34] Thus, GLD-1 appears to co-regulate functionally redundant subsets of the chitin binding domain gene family, the *puf* gene family and *oma-1/-2*.

The data accumulated to this point clearly shows that GLD-1 regulates the expression of its mRNA targets at the same level of RNA metabolism, by acting as a translational repressor. In addition, GLD-1 appears to have a very similar function in closely related nematodes *C. briggsae* and *C. remeini*. Loss of GLD-1 function results in essentially the same tumor phenotype in germ cells undergoing oogenesis in *C. briggsae*.[35] The orthologous *C. briggsae* and *C. remeini* GLD-1 proteins also repress the translation of *C. briggsae* and *C. remeini rme-2* mRNAs.[26,35] These data suggest that *C. briggsae* and *C. remeini* GLD-1 proteins likely repress the translation of similar sets of mRNA targets to *C. elegans* GLD-1.

mRNA TARGETS: FURTHER INSIGHTS INTO GLD-1 FUNCTION IN GERMLINE DEVELOPMENT

The identification of multiple GLD-1 targets that have important functions during various aspects of germline development provides significant clues as to why GLD-1 regulation of these mRNA targets is necessary to maintain the integrity of germline development. The various germline functions of the targets also begin to explain how the germline tumor/teratoma arises in the absence of GLD-1.

GLD-1 mediated translational repression of at least two mRNA targets, *cye-1* (Cyclin E) and *lin-45* (RAF kinase), contributes to meiotic prophase progression and prevention of return to mitotic proliferation (thus germline tumor and teratoma formation)(Lee et al, unpublished).[11] Although the exact sequence of events that result in an exit from pachytene and a return to proliferation in *gld-1* null mutant germlines remains uncertain, some features are known. The synaptonemal complex mediated pairing of maternal and paternal homologs during meiotic prophase is lost in *gld-1* null germlines, resulting in 12 condensed univalent chromosomes instead of 6 bivalent chromosomes. These 12 univalent chromosomes organize in a mitotic prophase configuration where a bipolar mitotic spindle with astral microtubules forms and a mitotic division occurs prior to additional rounds of S-phase and M-phase.[1,11] When CYE-1 (or CDK-2) mRNA is knocked down by RNAi in *gld-1* null mutants, germ cells in the central pachytene region retain meiotic prophase-like morphology and fail to display mitotic prophase chromosome morphology, spindle formation or mitotic divisions.[11] Cyclin E (CYE-1)/CDK-2 complexes are well known for their functions in the G1 to S phase transition and centrosome duplication.[36] However, since the CYE-1 dependent events of DNA replication and centrosome duplication occur in meiotic S-phase before germ cells enter meiotic prophase, misexpression of CYE-1 in the pachytene stage germ cells of *gld-1* null germlines likely leads to deregulation of other functions that result in return to mitosis.

In *C. elegans*, the ERK MAP kinase signaling module, which includes LIN-45 RAF, MEK-2 MEK and MPK-1 ERK, has multiple functions during oogenesis including being required for progression of germ cells through pachytene.[37] In wild type adult hermaphrodite germlines, MPK-1 is activated in proximal pachytene, but not earlier in meiotic prophase.[37] However, in *gld-1* null hermaphrodites, MPK-1 is activated earlier in meiotic prophase, possibly due to inappropriate early translation of *lin-45* mRNA in the absence of GLD-1. Importantly, in *gld-1; lin-45*, as well as in *gld-1; mpk-1* null double mutant germlines, germ cells arrest in pachytene and do not return to mitotic proliferation (Lee et al, unpublished). Experiments have not yet addressed whether mis-expression of CYE-1 or LIN-45 in early meiotic prophase of the wild type germline is sufficient to induce pachytene stage germ cells to return to mitotic proliferation. It is likely that translational repression of additional GLD-1 targets, along with CYE-1 and LIN-45, is necessary for meiotic prophase progression/oogenesis and that mis-expression of multiple targets would be required to provide conditions favorable for exit from meiotic prophase and ectopic proliferation.

PAL-1, a Caudal ortholog, is a homeodomain transcription factor that is required for specification of muscle cell precursors in the embryo. GLD-1 mediated translational repression of *pal-1* is required,[31] in part, to prevent transdifferentiation of pachytene germ cells into muscle-like somatic cells.[10] RNAi knockdown of *pal-1* in sensitized *gld-1* null hermaphrodite germlines significantly suppresses muscle cell differentiation but does not affect the extent of transdifferentiation to neurons. Thus translational repression of additional GLD-1 mRNA targets is likely to be important for preventing transdifferentiation into other somatic cell types.

CEP-1 is the primordial p53 protein in *C. elegans* and functions as a transcription factor that is necessary for DNA damage induced apoptosis. Schumacher et al[12] proposed that GLD-1 mediated translational repression of *cep-1* functions to titrate levels of CEP-1 in late pachytene so that the extent of apoptosis is responsive to DNA damage signals. In wild-type and the *cep-1(op236)* mutant at the permissive temperature (20˚C), CEP-1 does not promote apoptosis in proximal pachytene cells in the absence of DNA damage. However, in *cep-1* at the restrictive temperature (25˚C), high levels of CEP-1 caused by failure of translational repression leads to increased apoptosis, even in the absence of DNA damage.

Germline sex determination in *C. elegans* is controlled by a complex network of more than 20 genes.[3] The GLD-1 function to promote the male fate during hermaphrodite spermatogenesis, as revealed by a haploinsufficiency, may be explained by the identification of *tra-2* mRNA as a target.[28] TRA-2, which promotes the female germline fate, needs to be translationally repressed by GLD-1 to allow spermatogenesis. Consistent with this proposal was the identification of *fog-2*, which also functions to promote the male fate in the hermaphrodite but not in the male germline, as a cofactor with GLD-1 in the translational repression of *tra-2* mRNA.[38] FOG-2 appears to function in the translational repression mechanism as it binds to GLD-1 but does not appear to contact the *tra-2* mRNA directly. FOG-2 has not been found to function in translational repression of other GLD-1 mRNA targets examined, suggesting that it is likely a target specific cofactor. LIN-45 Raf, along with MEK-2 and MPK-1, also function in germline sex determination, promoting the male germ cell fate in both hermaphrodites and males.[37] Translational repression of *lin-45* mRNA would thus promote female germline development. The opposite effects on the sex determination network caused by GLD-1 mediated translational repression of *tra-2* and *lin-45* mRNAs may explain the incomplete penetrance of the sexual transformation observed in *gld-1* null mutants as well as the dominant/antimorphic sex determination phenotypes observed for specific *gld-1* alleles.[1,8]

mRNA TARGETS: TOWARDS DEFINING THE GLD-1 RNA BINDING MOTIF AND MECHANISM OF TRANSLATIONAL REPRESSION

One important question is how GLD-1 represses translation of its mRNA targets. Two major mechanisms of translational repression have been uncovered—repression prior to/ at translation initiation and post-initiation repression. Examples of repression prior to/at translation initiation include certain *Xenopus* maternal mRNAs. For these mRNAs, 5' CAP dependent translation initiation requires the interaction of eIF4G and eIF4E. CPEB binds to a 3'UTR element and recruits Maskin and eIF4E, which are thought to circularize the mRNA and block translation initiation by masking the eIF4G binding site on eIF4E,[39] Examples of post-initiation repression are *C. elegans lin-14* and *lin-28* and *Drosophila nanos* mRNAs, which are found on polysomes but without stable translation products being detected.[40-42] Two studies have suggested that GLD-1 can repress translation by either mechanism: repression of the *tra-2* mRNA has been reported to occur prior to/at initiation,[43,44] while repression of *pal-1* mRNA has been reported to be post-initiation.[31] One possibility to explain the contradictory *tra-2* and *pal-1* results is that some GLD-1 targets are repressed prior to/at initiation while others are repressed post-initiation. With the multiple mRNA targets identified through the GLD-1 IP/microarray analysis that are more significantly enriched than *tra-2* and *pal-1*, this issue can be addressed using polysome profiling of multiple targets. At least three outcomes are possible: all GLD-1 targets are repressed at the level of prior to/ at initiation, or all during elongation or some targets are repressed at the level of prior to/at initiation while others during elongation, which would indicate that GLD-1 employs more than one mechanism to repress translation. Translational repression mechanisms can also employ micro RNAs (miRNAs). It is not currently known whether miRNAs bind to GLD-1/ mRNA target complexes and contribute to the translational repression activity of GLD-1. Interestingly, work by Zhang et al[45] identified GLD-1 as a binding partner of the GW182 protein AIN-2. AIN-2 has been shown to interact only with miRNA-specific Argonaute proteins ALG-1 and ALG-2 and therefore it regulates protein expression of miRNA target mRNAs but not miRNA biogenesis. Thus miRNAs and AIN-2 may well collaborate with GLD-1 to repress the translation of some, if not all, GLD-1 mRNA targets.

Another outstanding question that can be addressed with the multiple targets is how GLD-1 specifically recognizes different mRNAs in vivo. GLD-1 has at least 100 but probably less than 500 mRNA targets (Lee et al, unpublished data). Thus GLD-1 mRNA specificity should not be too tight or too loose. GLD-1 may recognize one or more sequences and/or structural features. In a well executed biochemical study, Ryder et al[46] used recombinant GLD-1 purified from *E. coli* and one GLD-1 target, the *tra-2* 3'UTR, to provide the first solid evidence that GLD-1 has sequence specific binding activity. Purified recombinant GLD-1 forms a homodimer in vitro in the absence of RNA. This homodimer binds to two sub-sites (bipartite sites) in a single 22-nucleotide TGE (*tra-2* and *GLI* element) of the *tra-2* 3'UTR, in which both sub-sites are required for the strong binding. The authors showed that one sub-site has a hexanucleotide consensus (UACU(C/A)A) while the other sub-site was not defined. Another study by Galarneau and Richard[47] also showed that bipartite RNA motifs are required for the high affinity binding of GLD-1 to RNA in vitro. They defined the motif as one full and at least a half site of essentially the same sequence as the hexanucleotide consensus. Interestingly, the hexanucleotide consensus identified in the *tra-2* 3'UTR is also present in the GLD-1 binding regions of other mRNA targets such as *rme-2*, *glp-1* and *cep-1*,[12,26,30] indicating that the hexanucleotide consensus is an important sequence feature for GLD-1 recognition. However, in vivo, it is not clear whether GLD-1 forms only a homodimer or different complexes with different mRNA targets. It is also unclear whether all GLD-1 binding in vivo requires the hexanucleotide consensus and whether the presence of hexanucleotide consensus alone is sufficient to predict specific GLD-1 binding. Some results suggest that, in addition to the hexanucleotide consensus, GLD-1 containing complexes from worm cytoplasmic extracts likely utilize additional features (sequences, structures, binding partners, etc) to achieve specificity. First, several GLD-1 binding regions ranging from 50-250 base-long have been identified from the various GLD-1 mRNA targets (Lee et al, unpublished). While most of these GLD-1 binding regions contain the hexanucleotide consensus supporting its importance in GLD-1 binding, a few binding regions do not contain this motif. Thus GLD-1 must interact with a different motif(s) in these regions. Second, the hexanucleotide consensus does not appear sufficient for GLD-1 binding. Ryder et al[46] proposed several de novo GLD-1 targets (*mes-4*, *cdc-25.1*, *peb-1* and *puf-8*) that have at least one hexanucleotide in their 3'UTRs. However, none was enriched in the GLD-1 IP/microarray analysis. They also proposed 20 de novo targets that have a relaxed hexanucleotide consensus in the 3'UTR ((U > G > C/A)A(C > A)U(A/C > U)A). Only one of them, *pie-1*, is specifically enriched in the GLD-1 IP. While it is unlikely that all GLD-1 targets would be identified by the GLD-1 IP/microarray analysis, these data indicate that the hexanucleotide consensus alone is not sufficient to predict specific GLD-1 binding. Third, wild-type GLD-1 binds *rme-2*, *tra-2*, *cep-1* and *gna-2* mRNAs with similar efficiency. However, we found that two GLD-1 mutant proteins, *q126* and *op236*, which contain missense changes in the conserved RNA binding domain, are disrupted in both translational repression in vivo and binding in cytosol extracts for *tra-2* and *cep-1* mRNAs, while other targets are unaffected.[12,26] The target-specific defects of two GLD-1 mutant proteins suggest that GLD-1 may interact with distinct mRNAs through different sub-domains or with different partners. One possibility is that more than one consensus sequence exists. Alternatively, other factors (proteins and/or miRNAs?) are likely important in binding specificity and translational repression. Interestingly, Lehmann-Blount and Williamson[25] proposed that missense mutations of GLD-1 that produce a variety of germline defects are probably not involved in RNA binding directly when their structure is compared to that of SF1 (the splicing branch point sequence). Thus the authors proposed that if GLD-1 and SF1 adopt similar structural folds, many missense mutations likely disrupt GLD-1 structure. This structural change may disrupt specific interactions with GLD-1 binding proteins, which can cause the defects in RNA binding, the alteration of binding specificity and/or the inability to repress the translation.

HOW IS GLD-1 EXPRESSION REGULATED?

As discussed earlier, the nonhomogenous distribution of GLD-1 throughout the hermaphrodite germline (Fig. 2A,B) is essential for GLD-1's function to produce healthy gametes, to prevent germline tumor formation and to maintain germ cell identity. In the distal mitotic region, *gld-1* functions redundantly with *gld-2* (and *gld-3*), to inhibit the germline proliferative fate and/or promotes initiation of meiotic development. The low level of GLD-1 in the distal mitotic region and the steep rise of GLD-1 as germ cells enter meiosis indicate that it likely inhibits the translation of factors that promote the proliferation and/or inhibit the entry into meiosis. Since *gld-1* mRNA is present throughout the hermaphrodite germline, the level of GLD-1 protein is likely regulated at the level of translation and post-translation (see Fig. 3).[9] Indeed, at the distal end, FBF-1 and FBF-2, *puf* (Pumilio and FBF) family of RNA binding proteins, function redundantly to maintain the low level of GLD-1 protein.[48,49] FBF-1 and FBF-2 (collectively referred to as FBFs) have been shown to bind to FBF binding elements (FBE) in the *gld-1* 3'UTR and repress translation, thus inhibiting the accumulation of GLD-1.[49] The maintenance of low levels of GLD-1 at the distal end by FBFs is important to maintain the germline proliferative

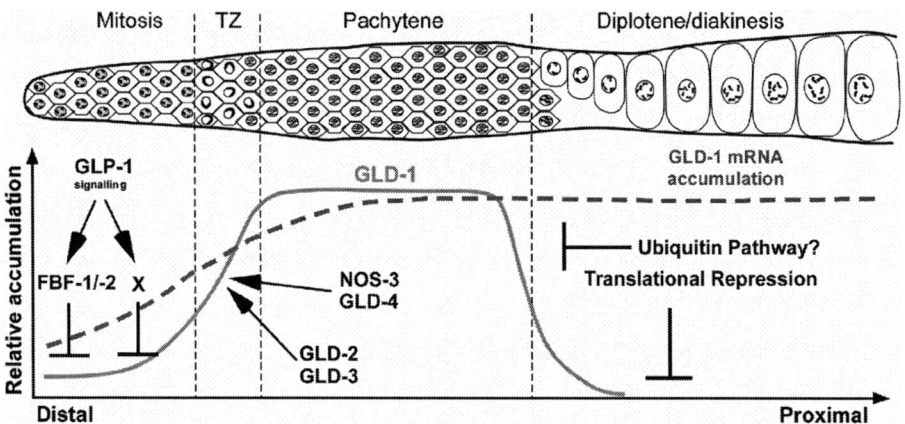

Figure 3. GLD-1 expression is regulated throughout the germline development by multiple pathways. The upper panel is a diagram of a single wild-type adult hermaphrodite gonad arm drawn linearly. The lower panel is a graph showing the relative accumulation pattern of *gld-1* mRNA (grey dashed line) and GLD-1 protein (grey [green] solid line) in adult hermaphrodites germline. The factors controlling GLD-1 accumulation are shown with inhibitory activities depicted by a bar and promoting activities depicted with an arrow. GLP-1 Notch signaling inhibits GLD-1 accumulation, partly through FBF-1/-2 and partly through an unknown factor(s).[17,62] Thus GLD-1 levels are low in the distal end where GLP-1 Notch signaling is highest. GLD-1 levels become higher as germ cells enter meiotic development in the transition zone where GLP-1 Notch signaling is reduced. Thus FBF is no longer inhibiting GLD-1 accumulation and the GLD-2/GLD-3 complex promotes GLD-1 accumulation redundantly with NOS-3. In addition, a cytoplasmic poly(A) polymerase, GLD-2 functions redundantly with another cytoplasmic poly(A) polymerase, GLD-4 to promote GLD-1 accumulation. NOS-3 and GLD-4 may function in the same pathway, although this has not yet been examined. GLD-1 sharply decreases as germ cells leave the pachytene stage to enter diplotene/diakinesis in the proximal region. Destabilization of GLD-1 and inhibition of new synthesis together are likely responsible for the absence of GLD-1 in the proximal region. A color version of this image is available at www.landesbioscience.com/curie.

fate.[49] Interestingly, FBFs also bind to the *gld-3* 3′UTR to repress the accumulation of GLD-3.[14] GLD-3 is an RNA binding protein, a Bicaudal-C homolog, that forms a complex with GLD-2, a cytoplasmic poly(A) polymerase. Thus by repressing translation of *gld-1* and *gld-3*, FBFs can strongly block initiation of meiotic development. However, it should be noted that additional regulators must also be involved as FBF is not required for the proliferative fate in larvae or in adults grown at 25˙C and, GLD-3 levels appear not to be significantly regulated in the adult (see Fig. 3).[17]

The steep increase in GLD-1 level at the end of the proliferative region and into the transition zone promotes initiation of meiotic development.[15,49] This steep increase is likely a result of a relief from FBF translational repression and a translational activation by GLD-2 and NOS-3. The level of FBFs is decreasing as germ cells enter meiosis,[50] which should then relieve the translation repression of *gld-1* and *gld-3* mRNAs. Thus GLD-1 and GLD-3 protein production should increase. Since GLD-3 can antagonize FBF RNA binding activity,[14,51] the increased production of GLD-3 can further antagonize FBF activity ensuring the complete relief of the translational repression of *gld-1* and *gld-3* mRNAs by FBF. The removal of both *gld-2* and *nos-3* activity results in a very significant reduction of GLD-1 accumulation,[15] suggesting that GLD-2 and NOS-3 activate the translation of *gld-1* mRNA independently. In fact, *gld-1* mRNA is a direct target of the

GLD-2 cytoplasmic poly(A) polymerase,[52] indicating that GLD-2 will increase the length of the *gld-1* mRNA poly(A) tail to enhance its translation. However, GLD-2 cannot be a sole activator of *gld-1* mRNA poly(A) addition (thus translation) since GLD-1 translation is still activated in the absence of GLD-2. Interestingly, another cytoplasmic poly(A) polymerase, GLD-4 has been identified recently.[53] The elimination of *gld-2* and *gld-4* activity together generates several germ cell defects that resemble those observed in *gld-1* null germlines and prevents GLD-1 accumulation,[53] suggesting that GLD-2 and GLD-4 act redundantly to add poly(A) tail to *gld-1* mRNA to activate translation (Fig. 3).

The level of GLD-1 protein sharply decreases as germ cells leave the pachytene stage to enter diplotene (Figs. 2 and 3) and is essentially absent in the late stage diakinesis oocytes in the proximal region. Apparently this is due to destabilization of GLD-1 as well as inhibition of new GLD-1 synthesis. These controls should be critical for the production of normal, healthy oocytes, however, the molecular mechanisms are not yet understood. Some limited and indirect data suggest that the ubiquitin mediated protein degradation is likely important for the destabilization of GLD-1 (Nayak et al, unpublished).

ASD-2, ANOTHER *C. ELEGANS* STAR PROTEIN, FUNCTIONS IN ALTERNATIVE SPLICING

Recently, another STAR protein, ASD-2 (Alternative Splicing Defective-2) has been identified genetically while searching for factors that control developmental switching of the mutually exclusive alternative splicing of the *let-2* gene during larval development.[54] *let-2* encodes a Type IV collagen and employs mutually exclusive exon 9 and exon 10 in body wall muscles. In embryos, an mRNA isoform with exon 9 is exclusively present, while in late larval and adult stages, an mRNA isoform with exon 10 predominates.[55,56] This suggests that switching of exon 9 and exon 10 likely alters the characteristics of basement membranes during larval development.

asd-2 mutant animals are defective in switching of alternative splicing, suggesting that ASD-2 promotes biased inclusion of exon 10 in the late larval and adult stages. Several missense mutations that disrupt the in vivo function of ASD-2 were identified and they are exclusively located in the STAR domain, indicating the importance of the STAR domain in its function.[54] ASD-2 is more closely related to GLD-1, Quaking and How/Who subfamily than to SF1 or SAM68 subfamilies. Thus ASD-2 may bind to sequences similar to the hexanucleotide consensus that is bound by GLD-1. In fact, a similar sequence is present in *let-2* intron 10 and, when mutated, alternative splice switching did not occur. ASD-2 also directly binds to this sequence in vitro, suggesting that *let-2* mRNA is likely a direct target of ASD-2.[54] It will be interesting to determine whether ASD-2 regulates alternative splicing of other genes with exon switching from early to late larval/adult stages.

CONCLUSION

The *C. elegans* STAR proteins GLD-1 and ASD-2 were first identified genetically. Germline development becomes defective upon loss of *gld-1* function while developmental switching of mutually exclusive alternative splicing of the *let-2* gene becomes defective with the removal of *asd-2* function. The identification of multiple mRNA targets make it possible to acquire a comprehensive understanding of how GLD-1 regulates its targets to control *C. elegans* germline development. With the identification of multiple mRNA targets, one can determine why GLD-1 regulation of specific targets is crucial in maintaining intact germline development. In addition, several key questions that have not been clearly addressed with STAR or other RNA binding proteins can be addressed; whether GLD-1 represses translation of all targets by the same mechanism or different sets of targets by distinct mechanisms and what are the specific features (sequences, structures, etc) and rules that distinguish targets from nontargets. GLD-1 may use distinct interacting partners such as different proteins and/or miRNAs on different sets of targets, which can potentially allow GLD-1 to have distinct sets of binding specificity while using different mechanisms to repress translation.

ACKNOWLEDGEMENTS

Work in the authors laboratories was supported by NIH GM63310 (TS) and American Cancer Society grant RSG0718101DDC (MHL). We are grateful to Swathi Arur and Dave Hansen for critically reading this chapter.

REFERENCES

1. Francis R, Barton MK, Kimble J et al. gld-1, a tumor suppressor gene required for oocyte development in Caenorhabditis elegans. Genetics 1995a; 139:579-606.
2. Zarkower D. Somatic sex determination. WormBook 2006; 1-12.
3. Ellis R, Schedl T. Sex determination in the germ line. WormBook 2007; 1-13.
4. Gibert MA, Starck J, Beguet B. Role of the gonad cytoplasmic core during oogenesis of the nematode Caenorhabditis elegans. Biol Cell 1984; 50(1):77-85.
5. Wolke U, Jezuit EA, Priess JR. Actin-dependent cytoplasmic streaming in C. elegans oogenesis. Development 2007; 134(12):2227-2236.
6. Gumienny TL, Lambie E, Hartwieg E et al. Genetic control of programmed cell death in the Caenorhabditis elegans hermaphrodite germline. Development 1999; 126(5):1011-1022.
7. Gartner A, Milstein S, Ahmed S et al. A conserved checkpoint pathway mediates DNA damage—induced apoptosis and cell cycle arrest in C. elegans. Mol Cell 2000; 5(3):435-443.
8. Francis R, Maine E, Schedl T. Analysis of multiple roles of gld-1 in germline development: interactions with the sex determination cascade and the glp-1 signaling pathway. Genetics 1995b; 139:607-630.
9. Jones AR, Francis R, Schedl T. GLD-1, a cytoplasmic protein essential for oocyte differentiation, shows stage-and sex-specific expression during Caenorhabditis elegans germline development. Dev Biol 1996; 180:165-183.
10. Ciosk R, DePalma M, Priess JR. Translational regulators maintain totipotency in the Caenorhabditis elegans germline. Science 2006; 311(5762):851-853.
11. Biedermann B, Wright J, Senften M et al. Translational repression of cyclin E prevents precocious mitosis and embryonic gene activation during C. elegans meiosis. Developmental Cell 2009; 17(3):355-364.
12. Schumacher B, Hanazawa M, Lee MH et al. Translational repression of C. elegans p53 by GLD-1 regulates DNA damage-induced apoptosis. Cell 2005; 120(3):357-368.
13. Kadyk LC, Kimble J. Genetic regulation of entry into meiosis in Caenorhabditis elegans. Development 1998; 125(10):1803-1813.
14. Eckmann CR, Crittenden SL, Suh N et al. GLD-3 and control of the mitosis/meiosis decision in the germline of Caenorhabditis elegans. Genetics 2004; 168(1):147-160.
15. Hansen D, Wilson-Berry L, Dang T et al. Control of the proliferation versus meiotic development decision in the C. elegans germline through regulation of GLD-1 protein accumulation. Development 2004a; 131(1):93-104.
16. Austin J, Kimble J. glp-1 is required in the germ line for regulation of the decision between mitosis and meiosis in C. elegans. Cell 1987; 51(4):589-599.

17. Hansen D, Schedl T. The regulatory network controlling the proliferation-meiotic entry decision in the Caenorhabditis elegans germ line. Curr Top Dev Biol 2006; 76:185-215.
18. Kimble J, Crittenden SL. Controls of germline stem cells, entry into meiosis and the sperm/oocyte decision in Caenorhabditis elegans. Annu Rev Cell Dev Biol 2007; 23:405-433.
19. Jones AR, Schedl T. Mutations in gld-1, a female germ cell-specific tumor suppressor gene in Caenorhabditis elegans, affect a conserved domain also found in Src-associated protein Sam68. Genes Dev 1995; 9:1491-1504.
20. Gibson TJ, Rice PM, Thompson JD et al. KH domains within the FMR1 sequence suggest that fragile X syndrome stems from a defect in RNA metabolism. Trends Biochem Sci 1993; 18(9):331-333.
21. Musco G, Stier G, Joseph C et al. Three-dimensional structure and stability of the KH domain: molecular insights into the fragile X syndrome. Cell 1996; 85(2):237-245.
22. Lewis HA, Musunuru K, Jensen KB et al. Sequence-specific RNA binding by a Nova KH domain: implications for paraneoplastic disease and the fragile X syndrome. Cell 2000; 100(3):323-332.
23. Di Fruscio M, Chen T, Bonyadi S et al. The identification of two Drosophila KH domain proteins: KEP1 and SAM are members of the Sam68 family of GSG domain proteins. J Biol Chem 1998; 273:30122-30130.
24. Vernet C, Artzt K. STAR, a gene family involved in signal transduction and activation of RNA. Trends Genet 1997; 13(12):479-484.
25. Lehmann-Blount KA, Williamson JR. Shape-specific nucleotide binding of single-stranded RNA by the GLD-1 STAR domain. J Mol Biol 2005; 346(1):91-104.
26. Lee MH, Schedl T. Identification of in vivo mRNA targets of GLD-1, a maxi-KH motif containing protein required for C. elegans germ cell development. Genes Dev 2001; 15(18):2408-2420.
27. Lee MH, Schedl T. Translation repression by GLD-1 protects its mRNA targets from nonsense-mediated mRNA decay in C. elegans. Genes Dev 2004; 18(9):1047-1059.
28. Jan E, Motzny CK, Graves LE et al. The STAR protein, GLD-1, is a translational regulator of sexual identity in C. elegans. EMBO J 1999; 18:258-269.
29. Xu L, Paulsen J, Yoo Y et al. Caenorhabditis elegans MES-3 is a target of GLD-1 and functions epigenetically in germline development. Genetics 2001; 159(3):1007-1017.
30. Marin VA, Evans TC. Translational repression of a C. elegans Notch mRNA by the STAR/KH domain protein GLD-1. Development 2003; 130(12):2623-2632.
31. Mootz D, Ho DM, Hunter CP. The STAR/Maxi-KH domain protein GLD-1 mediates a developmental switch in the translational control of C. elegans PAL-1. Development 2004; 131(14):3263-3272.
32. Evans TC, Hunter CP. Translational control of maternal RNAs. WormBook 2005; 1-11.
33. Lublin AL, Evans TC. The RNA-binding proteins PUF-5, PUF-6 and PUF-7 reveal multiple systems for maternal mRNA regulation during C. elegans oogenesis. Dev Biol 2007; 303(2):635-649.
34. Detwiler MR, Reuben M, Li X et al. Two zinc finger proteins, OMA-1 and OMA-2, are redundantly required for oocyte maturation in C. elegans. Dev Cell 2001; 1(2):187-199.
35. Nayak S, Goree J, Schedl T. fog-2 and the evolution of self-fertile hermaphroditism in Caenorhabditis. PLoS Biol 2005; 3(1):e6.
36. Hwang HC, Clurman BE. Cyclin E in normal and neoplastic cell cycles. Oncogene 2005; 24(17):2776-2786.
37. Lee MH, Ohmachi M, Arur S et al. Multiple functions and dynamic activation of MPK-1 extracellular signal-regulated kinase signaling in Caenorhabditis elegans germline development. Genetics 2007; 177(4):2039-2062.
38. Clifford R, Lee MH, Nayak S et al. FOG-2, a novel F-box containing protein, associates with the GLD-1 RNA binding protein and directs male sex determination in the C. elegans hermaphrodite germline. Development 2000; 127(24):5265-5276.
39. Stebbins-Boaz B, Cao Q, de Moor CH et al. Maskin is a CPEB-associated factor that transiently interacts with eIF-4E. Mol Cell 1999; 4(6):1017-1027.
40. Olsen PH, Ambros V. The lin-4 regulatory RNA controls developmental timing in Caenorhabditis elegans by blocking LIN-14 protein synthesis after the initiation of translation. Dev Biol 1999; 216(2):671-680.
41. Clark IE, Wyckoff D, Gavis ER. Synthesis of the posterior determinant Nanos is spatially restricted by a novel co-translational regulatory mechanism. Curr Biol 2000; 10(20):1311-1314.
42. Seggerson K, Tang L, Moss EG. Two genetic circuits repress the Caenorhabditis elegans heterochronic gene lin-28 after translation initiation. Dev Biol 2002; 243(2):215-225.
43. Goodwin EB, Okkema PG, Evans TC et al. Translational regulation of tra-2 by its 3' untranslated region controls sexual identity in C. elegans. Cell 1993; 75(2):329-339.
44. Kuersten S, Goodwin EB. The power of the 3' UTR: translational control and development. Nat Rev Genet 2003; 4(8):626-637.

45. Zhang L, Ding L, Cheung TH et al. Systematic identification of C. elegans miRISC proteins, miRNAs and mRNA targets by their interactions with GW182 proteins AIN-1 and AIN-2. Mol Cell 2007; 28(4):598-613.
46. Ryder SP, Frater LA, Abramovitz DL et al. RNA target specificity of the STAR/GSG domain post-transcriptional regulatory protein GLD-1. Nat Struct Mol Biol 2004; 11(1):20-28.
47. Galarneau A, Richard S. The STAR RNA binding proteins GLD-1, QKI, SAM68 and SLM-2 bind bipartite RNA motifs. BMC Mol Biol 2009; 10:47.
48. Zhang B, Gallegos M, Puoti A et al. A conserved RNA-binding protein that regulates sexual fates in the C. elegans hermaphrodite germ line. Nature 1997; 390(6659):477-484.
49. Crittenden SL, Bernstein DS, Bachorik JL et al. A conserved RNA-binding protein controls germline stem cells in Caenorhabditis elegans. Nature 2002; 417(6889):660-663.
50. Lamont LB, Crittenden SL, Bernstein D et al. FBF-1 and FBF-2 regulate the size of the mitotic region in the C. elegans germline. Dev Cell 2004; 7(5):697-707.
51. Eckmann CR, Kraemer B, Wickens M et al. GLD-3, a bicaudal-C homolog that inhibits FBF to control germline sex determination in C. elegans. Dev Cell 2002; 3(5):697-710.
52. Suh N, Jedamzik B, Eckmann CR et al. The GLD-2 poly(A) polymerase activates gld-1 mRNA in the Caenorhabditis elegans germ line. Proc Natl Acad Sci USA 2006; 103(41):15108-15112.
53. Schmid M, Kuchler B, Eckmann CR. Two conserved regulatory cytoplasmic poly(A) polymerases, GLD-4 and GLD-2, regulate meiotic progression in C. elegans. Genes Dev 2009; 23(7):824-836.
54. Ohno G, Hagiwara M, Kuroyanagi H. STAR family RNA-binding protein ASD-2 regulates developmental switching of mutually exclusive alternative splicing in vivo. Genes Dev 2008; 22(3):360-374.
55. Sibley MH, Johnson JJ, Mello CC et al. Genetic identification, sequence and alternative splicing of the Caenorhabditis elegans alpha 2(IV) collagen gene. J Cell Biol 1993; 123(1):255-264.
56. Graham PL, Johnson JJ, Wang S et al. Type IV collagen is detectable in most, but not all, basement membranes of Caenorhabditis elegans and assembles on tissues that do not express it. J Cell Biol 1997; 137(5):1171-1183.
57. Grant B, Hirsh D. Receptor-mediated endocytosis in the Caenorhabditis elegans oocyte. Mol Biol Cell 1999; 10(12):4311-4326.
58. Derry WB, Putzke AP, Rothman JH. Caenorhabditis elegans p53: role in apoptosis, meiosis and stress resistance. Science 2001; 294(5542):591-595.
59. Shimada M, Kawahara H, Doi H. Novel family of CCCH-type zinc-finger proteins, MOE-1, -2 and -3, participates in C. elegans oocyte maturation. Genes Cells 2002; 7(9):933-947.
60. Johnston WL, Krizus A, Dennis JW. The eggshell is required for meiotic fidelity, polar-body extrusion and polarization of the C. elegans embryo. BMC Biol 2006; 4:35.
61. Olson SK, Bishop JR, Yates JR et al. Identification of novel chondroitin proteoglycans in Caenorhabditis elegans: embryonic cell division depends on CPG-1 and CPG-2. J Cell Biol 2006; 173(6):985-994.
62. Hansen D, Hubbard EJ, Schedl T. Multi-pathway control of the proliferation versus meiotic development decision in the Caenorhabditis elegans germline. Dev Biol 2004b; 268(2):342-357.

CHAPTER 9

THE BRANCHPOINT BINDING PROTEIN
In and Out of the Spliceosome Cycle

Brian C. Rymond*

Abstract: The *Saccharomyces cerevisiae* branchpoint binding protein (BBP) is a 53 kDa pre-mRNA processing factor with characteristic STAR/GSG protein organization. This includes a central RNA binding site composed of an extended Type I KH domain with an adjacent QUA2 motif. Downstream of KH-QUA2 are two CCHC-type zinc knuckles and a proline-rich C-terminal interaction domain (Fig. 1A). The QUA1 homodimerization motif found upstream of the KH-QUA2 sequence in other STAR/GSG family members is absent in BBP and replaced by a site for the phylogenetically conserved binding partner, Mud2/U2AF65. BBP's name reflects the fact that it binds the conserved RNA sequence, UACUAAC, called the branchpoint motif found near the 3' end of yeast introns. This sequence contains the catalytic adenosine (underlined) which directs the first RNA transesterification reaction in splicing chemistry. BBP recruitment to the branchpoint initiates a series of spliceosomal subunit addition and rearrangement events that ultimately configures the active site of this enzyme.[1] The mammalian homolog, ZFM1/ZNF162/D11S636/SF1 (henceforth, SF1), was first identified in a screen for genes associated with Type 1 multiple endocrine neoplasia[2] and was subsequently shown to act similarly to BBP in mammalian splicing.[3,4] BBP/SF1 is essential for viability in organisms spanning the evolutionary spectrum from yeast to *Caenorhabditis elegans* to mice. In addition, mice heterozygous for a *SF1* knockout allele show enhanced susceptibility to azoxymethane-induced colon tumorigenesis[5] adding BBP/SF1 to the growing list of RNA processing factors implicated in genetic disease.[6] Summarized below is our current understanding of BBP structure and its proposed multifaceted contribution to mRNA biogenesis and function. Reference to SF1 will be made to fill gaps in our understanding of BBP and to highlight areas of clear similarity or difference between yeast and mammals.

*Brian C. Rymond—Biology Department, University of Kentucky, 675 Rose Street, Lexington, Kentucky 40506-0225 USA. Email: rymond@uky.edu

Post-Transcriptional Regulation by STAR Proteins: Control of RNA Metabolism in Development and Disease, edited by Talila Volk and Karen Artzt.
©2010 Landes Bioscience and Springer Science+Business Media.

BBP AND SF1 ARE SITE-SPECIFIC RNA BINDING PROTEINS

MSL5, the gene encoding BBP, was one of seven *MUD2 Synthetic Lethal* mutants identified by the Rosbash lab as inviable when combined with a nonlethal *mud2* mutation.[7] Mud2 is the yeast homolog of the mammalian U2 small nuclear ribonucleoprotein (snRNP) particle recruitment factor U2AF65.[8] *MSL5* is essential in yeast and *msl5* mutants are splicing impaired especially with pre-mRNA that has a nonconsensus branchpoint sequence.[9,10] Extracts depleted of BBP fail to assemble the so-called CC2 commitment complex that corresponds to the first stable interaction between proteins bound at the pre-mRNA branchpoint and the U1 snRNP bound at the 5' splice site.[7,11] Commitment complex (or mammalian E-complex)[12,13] assembly confers a competitive advantage to the RNA during in vitro splicing and, in vivo, is thought to direct (or commit) pre-mRNA into the splicing pathway and away from a presumed default path of RNA export from the nucleus.

BBP and SF1 bind directly and specifically to the branchpoint motif of intron-bearing RNA.[14,15] Indeed, the most prevalent natural yeast branchpoint sequence, UACUAAC, was found as the most preferred target of BBP by SELEX.[15] The SELEX studies also suggest that less conserved flanking nucleotides and in particular an adjacent upstream hairpin structure may enhance BBP binding.[15] RNA secondary structures can increase or decrease splicing efficiency[16] and it is conceivable that a branchpoint-proximal hairpin, while not a general feature of yeast introns, can influence the processing of select pre-mRNAs through stabilized BBP association.

The SF1/BBP RNA binding site is organized into a $\beta\alpha\alpha\beta\beta\alpha\alpha$ protein folding pattern typical of an extended Type 1 KH domain.[17] Based on an NMR structure of a SF1 peptide bound with a synthetic oligonucleotide,[18] the splicing substrate is expected to lie within a cleft formed by the conserved GXXG loop between the second and third α helices and a highly variable loop between the second and third β strands (Fig. 1B). As expected, mutations within KH-QUA2 can greatly reduce the strength and specificity of RNA binding.[14,15,18,19] In addition, mutational studies conducted with recombinant peptides show that the first zinc knuckle segment modestly increases BBP's affinity for RNA.[15] Surprisingly, the zinc knuckle per se is not required as mutation of the first two metal coordinating residues has little impact on RNA binding. This zinc knuckle is predicted to interact with nonconserved intron sequence upstream of the branchpoint in SF1[18] and in BBP feature also appears to bind RNA nonspecifically.[15] A second zinc knuckle is not conserved in SF1 and its function in BBP is more ambiguous as its deletion enhances rather than reduces RNA interaction.[15] This unexpected finding has prompted speculation that the second zinc knuckle may facilitate later steps in spliceosome assembly when release of BBP from the branchpoint is required.[15]

SF1 binds RNA approximately 200-fold less tightly and with less specificity than its yeast counterpart[20] but can be made much more BBP-like by R240K and K241R substitutions within alpha helix 4 of QUA2,[15,19,20] (see Fig. 1B). K241R is predicted to favor stacking interactions between the arginine residue of SF1 and the branchpoint adenosine; it is less clear why the R240K enhances SF1-RNA association.[18,19] The more limited affinity and specificity of SF1 toward RNA is understandable given the greater natural variability of the mammalian branchpoint motif, YNCURAY.[21] Weaker binding by SF1 favors competition among sub-optimal splice sites and fosters alternative splicing so critical for mammalian gene expression.[1,22] In contrast, only ~5% of yeast genes have introns and most of these contain a single unregulated intron. Tight and specific binding by BBP is expected to favor the efficient constitutive splicing typical in yeast.

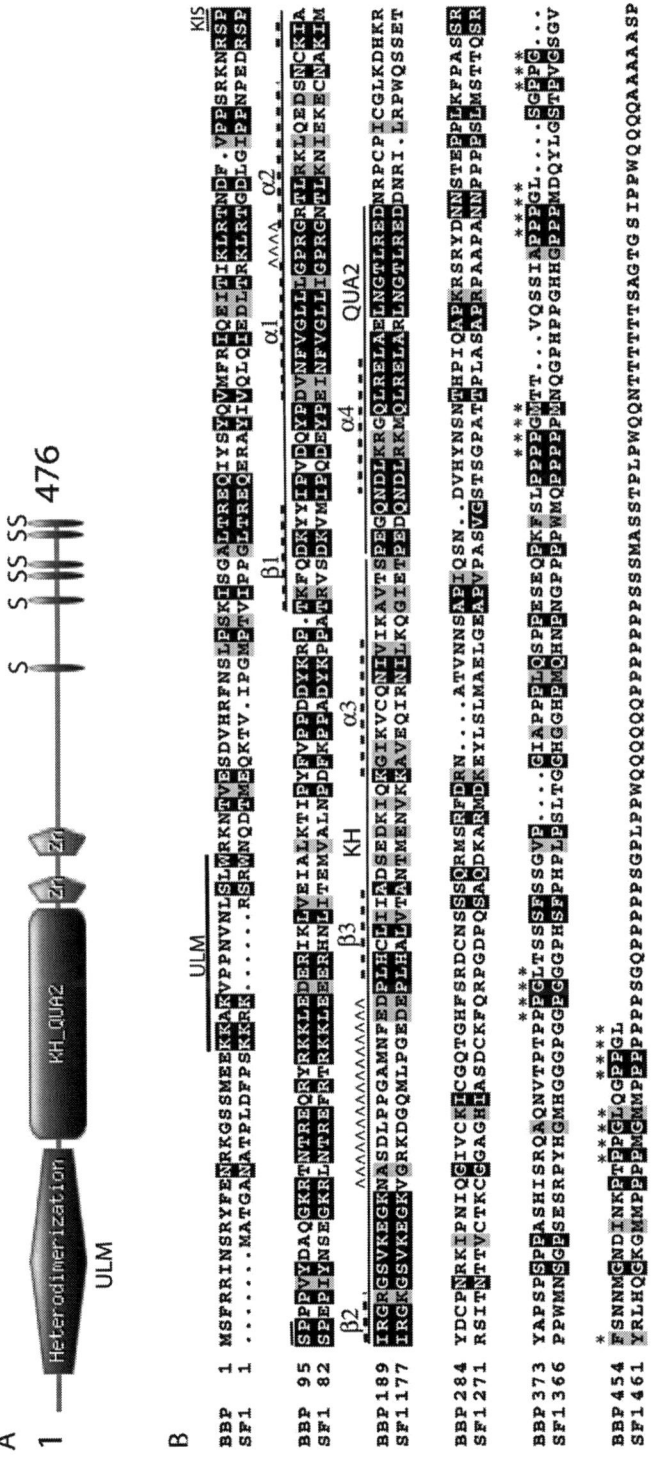

Figure 1. Domain organization of the BBP/SF1 splicing factor. A) Schematic representation of BBP. Indicated are the ULM domain-containing region that mediates Mud2 association (Heterodimerization); the RNA binding domain (KH_QUA2), two CCHC zinc knuckles (Zn) and the positions conforming to the Smy2 binding site, PPG{F/I/L/M/V}. The Smy2 sequences overlap a clathrin adaptor appendage domain that has not been scored for function in BBP. B) Alignment of the yeast BBP and human SF1 proteins using Blossom 62 matrix with identities shaded black and conserved residues shaded gray. Shown above BBP is the ULM sub-sequence critical to the SF1/U2AF65 interaction, the location of the KIS phosphorylation motif and the βααββαα secondary structures (thick dashed line) of the KH-QUA2 motif. Also shown are the two loop sequences that contribute to the RNA binding channel (^) and the six Smy2 binding sites (*).

A BBP-MUD2 HETERODIMER FUNCTIONS IN BRANCHPOINT RECOGNITION

The genetic incompatibility of certain *msl5* and *mud2* mutations reflects physically and functionally the interaction between the encoded proteins.[7,10,23,24] Unlike *MSL5*, *MUD2* is not essential and *mud2* null mutants grow well although with modestly reduced splicing efficiency.[25,26] BBP and Mud2 largely copurify from yeast as a simple heterodimer that does not require an RNA tether.[10] The amino terminus of BBP contains the Mud2 binding site; amino acids 41-141 are sufficient for stable Mud2 association[4] while deletion of amino acids 2-56 blocks recovery of the BBP-Mud2 heterodimer.[10] An atypical RNA recognition motif (RRM) called the U2AF homology motif (UHM) in the carboxyl terminus of Mud2 serves as the contact point for BBP.[4,24] The UHM is adapted to protein rather than RNA interaction by the presence of a C-terminal alpha helix that occludes the remnant RNA binding surface.[27] UHM elements bind a complementary surface within target proteins called the UHM ligand motif (ULM). Cooperative effects contributed by U2AF65 increase SF1's affinity for RNA ~ 20-fold.[24] U2AF65 itself binds RNA at a pyrimidine-rich sequence commonly positioned between the branchpoint and the 3' splice site of mammalian introns. This interaction is mediated by the two canonical RRM motifs positioned upstream of the U2AF65 UHM and is required for U2AF65 stimulated U2 snRNP recruitment in spliceosome assembly.[8,28-31]

A comparison of the BBP-Mud2 and SF1-U2AF65 interactions reveals both conserved and unique features of subunit association. For instance, similar to BBP the N-terminus of SF1 binds a hydrophobic pocket between helices A and B of the UHM domain of U2AF65.[32] The SF1 ULM includes residues 15KKRKRSRW22 where mutation of the conserved serine and tryptophan residues (underlined), phosphorylation of this serine or multiple charge reversals of the basic amino acids inhibit SF1-U2AF65 interaction.[33,34] The SF1-U2AF65 association is further stabilized by phosphorylation of SF1 79SPSP82 (Fig. 1B) by the KIS kinase, an enzyme composed of another UHM motif fused to a unique kinase domain. The putative ULM motif within BBP is interrupted by a seven amino acid insertion. While this specific sequence of BBP may augment Mud2 binding it is not required for this interaction.[4] The SPSP phosphorylation motif is conserved in position and sequence within BBP but it is not known whether this sequence (93SPSP96) or a second copy within BBP (376SPSP379) is modified or contributes to Mud2 association.

Mud2 binds RNA but has less conserved RRM sequences and lacks the N-terminal domain of clustered arginine-serine (RS) peptides that provide RNA substrate specificity for U2AF65. In addition, the polypyrimidine U2AF65 binding site is not conserved in yeast introns suggesting relaxed stringency for Mud2 association. Mud2 interaction with RNA requires its BBP binding site and this is consistent with BBP-directed substrate association. Indeed, crosslinking of Mud2 to RNA does not increase when Mud2 is overexpressed which suggests RNA association only in the context of the BBP-Mud2 heterodimer.[25,35] This reduced binding specificity correlates with a more limited role for Mud2 in splicing. Whereas U2AF65 is critical for efficient splicing in mammals, *MUD2* is not required for cell viability and the removal of Mud2 or all natural RNA sequence downstream of the branchpoint does not inhibit first step splicing catalysis.[25,36,37] The RNA binding site for Mud2 is unidentified, but extrapolating from U2AF65, likely consists of poorly conserved sequence near the branchpoint.

The mammalian U2AF65 protein is found stably associated with a smaller U2AF subunit called U2AF35.[27,38] This interaction occurs through a ULM motif within U2AF65 and a UHM within U2SF35. U2AF35 can be crosslinked to the pre-mRNA 3' splice site and this interaction stabilizes the association of U2AF65 with the adjacent polypyrimidine tract.[39-41] Together SF1, U2AF65 and U2AF35 complete a 3-subunit assemblage that defines the 3' end of the intron (i.e., 5' SF1/branchpoint ->U2AF65/polypyrimindine tract ->U2AF35/3' splice site). In the fission yeast, the large and small subunit of U2AF form a stable complex with the SF1 homolog[42] and presumably bind the splicing substrate as a unit. The mammalian SF1/U2AF65/U2AF35 proteins may also assemble an extra-spliceosomal complex[43] although the SF1 association is much more labile than what is seen in fission yeast. Homologs of U2AF35 are widely distributed in nature but are not found in baker's yeast.[10] The absence of a U2AF35-like subunit and the independence of first step catalysis from sequence downstream of the branchpoint are two of the clearest differences between yeast and mammals in the basal splicing apparatus.

BBP-MUD2 AND THE DYNAMICS OF EARLY SPLICEOSOME ASSEMBLY

Spliceosome assembly progresses through an ordered sequence of snRNP particle addition, rearrangement and dissociation steps that collectively define the spliceosome cycle (Fig. 2 and reviewed in ref. 1). Eight structurally related DExD/H-box protein ATPases serve to advance the steps of assembly and disassembly and assure the fidelity of splice site choice.[44,45] The stable binding of the initiating unit, the U1 snRNP particle, requires Watson-Crick basepairing between the 5' end of the U1 snRNA and the pre-mRNA 5'splice site. This interaction is stabilized by the yeast U1 C protein and other proteins associated with the U1 snRNP particle, the pre-mRNA cap-binding complex and the pre-mRNA branchpoint regions. The resulting commitment complex serves to retain the unprocessed pre-mRNA in the nucleus and to promote the next step of spliceosome assembly, namely the ATP-dependent addition of the U2 snRNP particle.[46-51] Two integral U1 snRNP proteins Prp40 and Prp39 bind BBP while another, Snu56, interacts with Mud2 and thereby help initially juxtapose the 5' splice site with the branchpoint motifs that direct in the first chemical step in splicing.[7,23,52] The U1 snRNP protein Prp39 was also reported to bind Mud2[23] but this interaction is indirect and mediated by the stable BBP-Mud2 heterodimer.[10]

The commitment complex is normally a transient intermediate easily seen only when the next step in spliceosome assembly is blocked by the removal of ATP or by the inactivation of an essential U2 snRNP component. Otherwise, the U2 snRNP particle is rapidly incorporated to form the prespliceosome. As the BBP-branchpoint interaction is incompatible with U2 snRNA basepairing across this same region of pre-mRNA, a major reorganization of the pre-mRNP must occur during prespliceosome formation.[24] The initial docking of the U2 snRNP particle is likely protein-directed. This step is perhaps better studied in mammals where U2AF6 binds SF3b155 (also called SAP155),[53] a conserved protein within the SF3b sub-particle of the U2 snRNP.[54-56] The U2AF65-SF3b155 interaction occurs though a ULM segment located between amino acids 317-357 of SF3b155 and the same UHM motif of U2AF65 that mediates the U2AF65-SF1 interaction. SF3b155 binding is expected to lower SF1's affinity for RNA.[32,57,58] SF3b155 itself can be crosslinked across both sides of the branchpoint when integrated into the splicing complex,[57] an observation consistent with both SF3b155 and its binding partner U2AF65 acting to stabilize the essential U2 snRNA/pre-mRNA branchpoint interaction.[31] Another UHM-bearing RNA-binding protein called Puf60 binds a second ULM within SF3b155

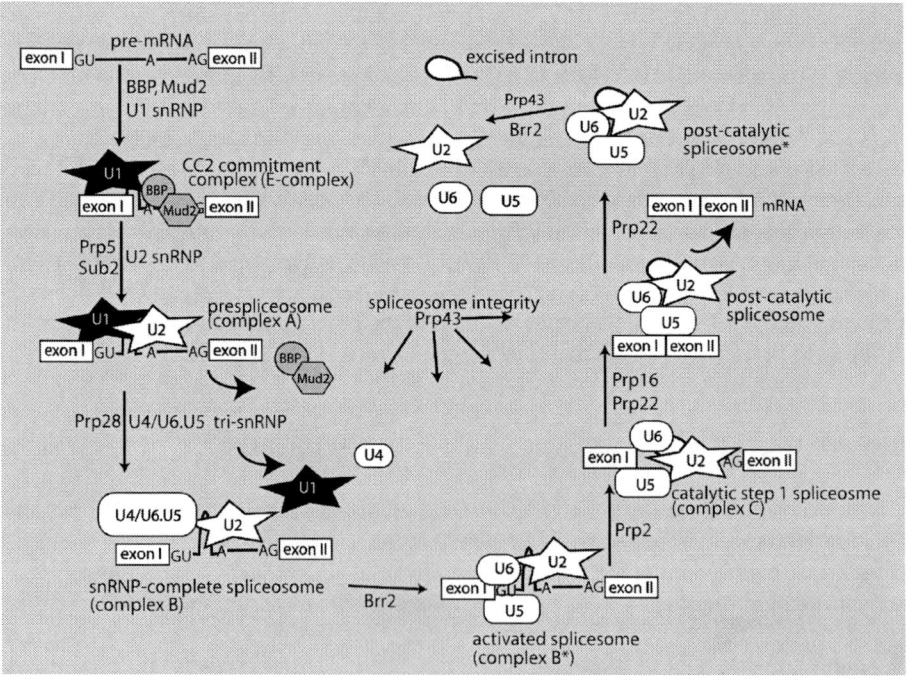

Figure 2. Schematic representation of the spliceosome cycle. A two-exon pre-mRNA is shown progressing through the steps of snRNP (U1, U2, U4/U6.U5) addition, rearrangement, splicing and subunit release. The assembly process is driven by eight DExD/H-box ATPases (Prp5, Sub2, Prp28, Prp2, Prp16, Prp22, Brr2 and Prp43) that promote conformational changes within the evolving RNP particle. *prp43* mutants suppress multiple assembly defects and Prp43p may function to dissociate defective spliceosomes as well as the postcatalytic spliceosome[158,159] (spliceosome integrity). An asterisk on the postcatalytic spliceosome shows the form of this complex typically recovered from yeast.[160,161]

(residues 194-229) and acts cooperatively with U2AF65 to enhance splicing on RNAs with weak 3' sites.[59]

The yeast SF3b155 homolog is Hsh155[60] and similar to the U2AF65- SF3b155 interaction, Hsh155 binds Mud2 (Fig. 3).[55] There is no Puf60 homolog in yeast although two-hybrid interactions between Mud2 and the U2 snRNP proteins Cus1 and Prp11 suggest additional stabilizing contacts for U2 snRNP particle recruitment[25,55] that are consistent with similar observations from the mammalian system.[57,61] Hsh155 also binds the conserved Prp5 DExD/H-box protein suggested to play a non-enzymatic role in bridging the U1 snRNP-U2 snRNP particles.[55,62] Prp5's enzymatic activity is also critical and helps configure the U2 snRNP particle into a state competent for spliceosome assembly[63] and assure the fidelity of splice site choice.[64,65]

Although docking of the U2 snRNP particle almost certainly weakens BBP's association with the RNA, its actual displacement likely requires a second DExD/H-box protein, Sub2 (Fig. 3). Sub2 and its mammalian counterpart, UAP56, function both in pre-mRNA splicing and in the export of mRNA from the nucleus.[66-68] Sub2 is recruited to the nascent transcript as part of the multi-subunit transcription and export (TREX) complex independent of whether an intron is present in the RNA.[66,69] *SUB2* is required for viability in most genetic backgrounds and yet deletion of *MUD2*[70] or mutation of the

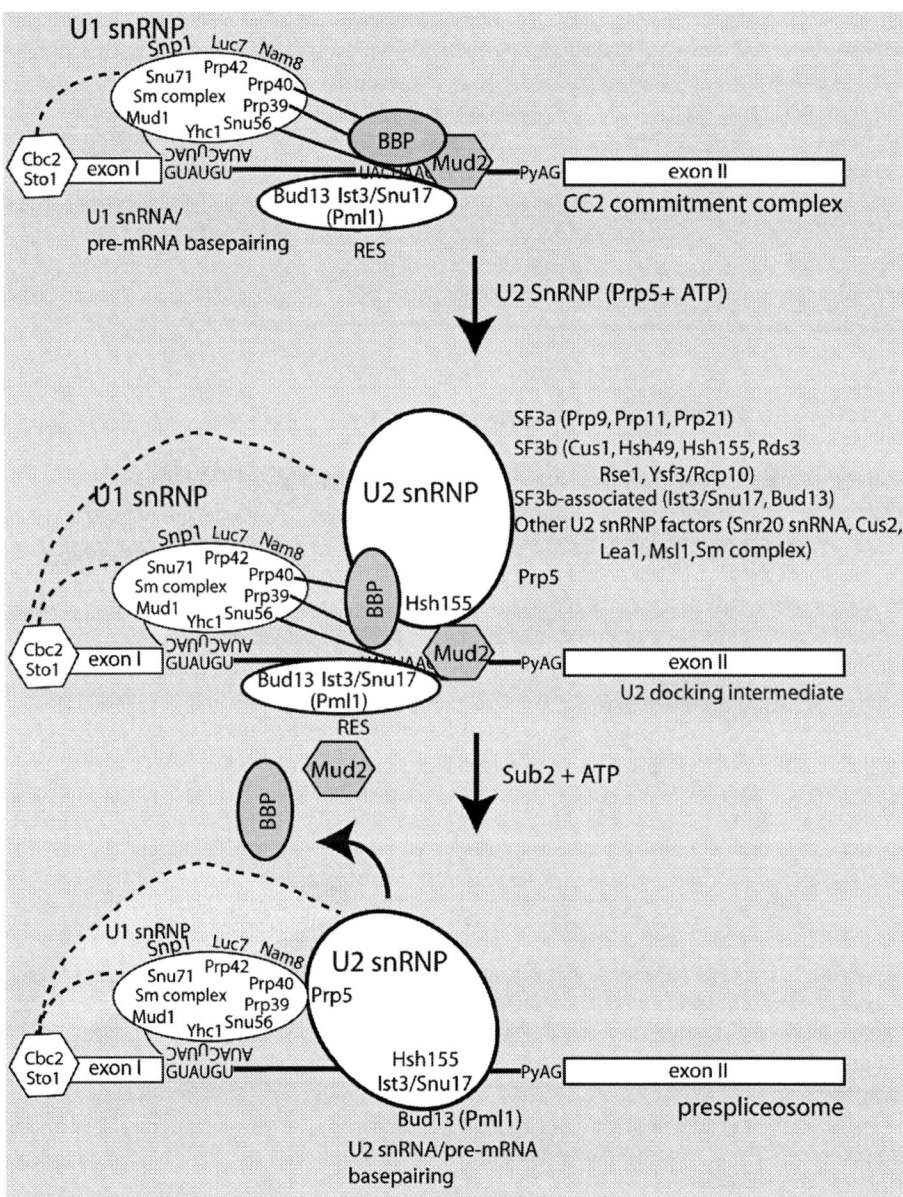

Figure 3. BBP-Mud2 displacement in prespliceosome formation. A hypothetical 2-step path for the conversion of the CC2 commitment complex to the prespliceosome is shown. The U1 snRNP-associated proteins are listed inside the oval or on the oval surface. The U1 snRNA is shown inverted and basepaired with the 5' splice site. Two-hybrid interactions reported for BBP and Mud2 with the U1 snRNP particle are shown in solid lines with biochemical or genetic interactions between the cap binding complex (Cbc2, Sto1) and the U1 and U2 snRNP particles shown by dashed lines. The U2 snRNP proteins are listed to the right of the U2 snRNP oval. Although shown as an early event, ATP hydrolysis by Prp5 may occur after U2 snRNA/pre-mRNA basepairing.[65] The association of RES with the branchpoint region is conjecture based on published genetic and biochemical observations.

Mud2 binding domain within BBP[10] eliminates this requirement. The suppression of *sub2* lethality by *mud2* and *msl5* mutations has been taken to suggest that the critical function of Sub2 is to advance spliceosome assembly by the removal of Mud2 and BBP.[10,70] Indeed, extracts with reduced Sub2 activity show spliceosome assembly defects consistent with improper U2 snRNP incorporation.[67,70,71] From this perspective, persistent BBP or Mud2 association with the splicing substrate restricts spliceosome assembly while either the presence of Sub2 or mutations that weaken BBP-Mud2 affinity for RNA bypass this block. Null alleles of *msl5* remain lethal independent of the *SUB2* status as BBP is required for cell viability whether or not the Sub2 helicase is present. In contrast to U2AF65, once the Sub2-dependent step has been executed and the prespliceosome is fully formed BBP and Mud2 no longer stably associate with the splicing apparatus.[37]

CO-TRANSCRIPTIONAL PRE-mRNA SPLICING

Research conducted over the past decade has provided much evidence for crosstalk among the pathways governing transcription, splicing, mRNA 3' end formation, nuclear export, translation and mRNA decay (reviewed in refs. 72,73). For instance, several groups have used cell-based assays to investigate the temporal and functional coupling of spliceosome assembly with transcription. The general approach taken was to induce gene expression and then monitor splicing factor or RNA polymerase association across a specific transcription unit. Target chromatin was recovered with splicing factor or RNA polymerase-specific antibodies and the sites of nucleic acid-protein interaction determined by quantitative PCR (see refs. 74-83 and references within). Three generalizations can be made from this work. First, in both yeast and metazoa splicing factor recruitment to the target gene requires transcriptional activation and RNA synthesis. Second, while spliceosomes can be assembled co-transcriptionally in both yeast and mammals, co-transcriptional splicing appears much more common in mammals. Third, the general pattern of spliceosome assembly observed in vitro (Fig. 2) is recapitulated in vivo. For instance, BBP, Mud2, U1 snRNP and the cap-binding complex factors are recruited early, U2 snRNP recruitment requires prior U1 snRNP association, U4/U5.U6 tri-snRNP addition occurs after prespliceosome assembly and the postcatalytic spliceosome accumulates only if tri-snRNP addition is allowed. This ordered addition of splicing factors is less compatible with an alternative model for the cellular spliceosome based on a preformed penta-snRNP particle,[84] a complex already demonstrated not to be essential for splicing in vitro.[85]

The carboxyl-terminal domain (CTD) of the largest RNA polymerase II subunit is required but is not sufficient to promote co-transcriptional pre-mRNA splicing (see refs. 86,87 and references within). The CTD is positioned adjacent to the exit channel for RNA on the polymerase, an ideal location to facilitate splicing factor recruitment and deposition.[88] In yeast, the U1 snRNP protein Prp40 binds the phosphorylated CTD[89] in addition to BBP.[7] In principle, this association of U1 snRNP (and indirectly BBP) with the RNA polymerase provides an efficient means to load the factors nucleating spliceosome assembly as soon as the pre-mRNA 5' splice site and branchpoint regions are transcribed. SF1 and a number of other splicing factors such as mammalian CA150 protein (which also binds the CTD and interacts with SF1), Cus2 (mammalian TAT-SF1), yeast Prp45 (mammalian NCoA-62/hSKIP) and certain mammalian SR protein family members appear to function in both splicing and transcription and may help functionally integrate these two processes.[72,90,91] Mammalian U1 snRNA also interacts with the TAF15 transcription factor[92] although the relationship, if any, between this complex and the splicing machinery is obscure. What is clear is that inhibiting the rate of transcriptional elongation by genetic or chemical means profoundly alters the pattern of splice site selection and mRNA 3' end processing (e.g., refs. 79,93,94). The potential contribution of transcriptional pause sites, elongation factors and nucleosome positioning or modification states to the regulation of pre-mRNA processing is a fascinating but largely undeveloped area in our understanding of gene expression.[72,73,95]

BUT IS BBP REALLY AN ESSENTIAL SPLICING FACTOR?

One of the more puzzling observations concerning BBP is that while inactivation of this protein inhibits cellular splicing its removal from extracts has little impact on splicing in vitro. BBP-depleted extracts fail to produce the gel-shift band defining the CC2 commitment complex but this is seen only when further spliceosome assembly is artificially blocked. Otherwise, the removal of >99% of BBP does not appreciable reduce the rate of spliceosome assembly or the extent of pre-mRNA splicing.[37] An argument has been made that BBP may act catalytically in splicing with only trace amounts needed for function[37] or to stimulate the recycling of other splicing factors. Direct evidence to support such models has not been forthcoming, however. As it stands the data support a simpler possibility, namely, that branchpoint recognition by BBP-Mud2 can be bypassed in vitro and the U2 snRNP recruited directly to the U1 snRNP-bound pre-mRNP.

If the BBP function can be so easily bypassed in spliceosome assembly then why is this protein necessary for cell viability? One possibility is that BBP is required only for the splicing of certain transcripts, such as pre-mRNAs with sub-optimal splice sites or other features that render these RNAs less efficient in spliceosome assembly. Substrate-specific differences have been observed after the inactivation of other core splicing factors in yeast (e.g., refs. 26,96). Also, SF1 knockdown experiments suggest that SF1 is not needed for the processing of at least certain mammalian pre-mRNAs.[97] Alternatively, BBP may be a general splicing factor but essential only for a feature of splicing or an aspect of pre-mRNA fate restricted to the cellular state. For instance, BBP may be critical for co-transcriptional RNA processing or where the nuclear surveillance of inefficiently processed RNA becomes relevant. In support of the latter possibility it is interesting to note that the growth of yeast mutant for *RRP6* is exacerbated when BBP becomes limiting.[98] Rrp6 is a nuclear exosome subunit[99] that contributes to the turnover of improperly processed RNA. The *msl5-rrp6* genetic interaction raises the possibility that BBP may mitigate the cytotoxic effect of improperly processed RNAs. Finally, it remains possible that the BBP requirement for cell viability reflects an unknown novel contribution of BBP to cellular biochemistry.

BBP IS NEEDED FOR THE NUCLEAR RETENTION
OF UNPROCESSED PRE-mRNA

Eukaryotic organisms prevent the export of unspliced RNA from the nucleus in a number of ways. For example, the prevalence of co-transcriptional splicing in mammals restricts most RNA to the site of synthesis until it is largely or completely processed. This may be less common in yeast where many pre-mRNAs appear to be spliced post-transcriptionally.[100] However in both systems early acting splicing factors restrict the export of pre-mRNA until splicing is complete. Mutation of either *MSL5* or *MUD2* significantly increases the abundance of unspliced pre-mRNA in the cytoplasm.[9,10,35] The reduction of splicing substrate through increased nuclear export of pre-mRNA almost certainly contributes to the slow growth of *msl5* mutants. Indeed, deletion of the gene for the Yra2 RNA export factor improves the growth of yeast limited for BBP suggesting that slowed RNA export partially compensates for the detrimental affects of reduced BBP levels.[98] Any pre-mRNA that successfully exits the nucleus is expected to compete with properly processed mRNA for ribosome occupancy and, if not degraded, has the potential to direct the synthesis of toxic peptides. This appears to be the case as a number of viable *msl5* mutants become lethal when the cytoplasmic nonsense mediated decay (NMD) pathway that degrades cytoplasmic pre-mRNA is blocked by mutation of *UPF1*.[9]

While an exhaustive survey has not been conducted, a number of other yeast splicing factors clearly contribute to the nuclear retention of unprocessed RNA. These include the U1 snRNA, the U2 snRNP particle proteins Prp9 and Ysf3/Rcp10, the U2 snRNP-associated proteins Bud13 and Ist3/Snu17 and the U4/U6.U5 tri-snRNP protein, Prp6.[35,55,56,101,102] Bud13 and Ist3/Snu17 are especially interesting as these proteins enhance the splicing of atypical pre-mRNAs such as those with a nonconsensus branchpoint sequence and Mer1-dependent meiotic pre-mRNA.[103,104] Ist3/Snu17 interacts with Mud2 in the two-hybrid assay[55] and is structurally quite similar to the mammalian SF3b14a protein that is found crosslinked to the pre-mRNA branchpoint after SF1 displacement.[105] While not essential, mutation of *IST3/SNU17* alters the electrophoretic mobility of the U2 snRNP particle suggesting a defect in particle assembly or stability.[106] Ist3/Snu17 and Bud13 are detected in the SF3b sub-particle of the U2 snRNP that includes the phylogenetically conserved proteins Hsh155, Rse1, Hsh49, Cus1 and Rds3 and Ysf3/Rcp10.[54-56] When the reciprocal purification is done, multiple SF3b proteins copurify with Ist3/Snu17.[55] Ist3/Snu17 and Bud13 can also be recovered with the Pml1 protein in a three-component assemblage called the RNA Export and Splicing (RES) complex.[56] Pml1 is needed for the efficient nuclear retention of pre-mRNA but this protein does not appear to act in splicing. The RES-SF3b association raises the possibility that the phylogenetically conserved RES complex helps mark unspliced pre-mRNA for nuclear retention soon after synthesis but hands this job off to the U2 snRNP later in assembly. At that time, at least Bud13 and Ist3/Snu17 remain bound to the splicing apparatus, possibly interacting with SF3b near the branchpoint to enhance splicing efficiency (Fig. 3).

The release of pre-mRNA from the nucleus observed after splicing factor inactivation is not restricted to yeast. For instance, when components of mammalian SF3b are inactivated pharmacologically or by RNAi splicing is inhibited and pre-mRNA is released into the cytoplasm.[107,108] Finally, it is also clear that proteins unrelated to splicing help retain unprocessed RNA in the nucleus[109-111] although little is known at a mechanistic level of how these factors function in pre-mRNA retention or communicate with the splicing apparatus.

UNCOUPLING PRE-mRNA SPLICING FROM THE SYNTHESIS
OF FUNCTIONAL mRNA

Inappropriately or incompletely processed RNA is targeted for destruction by nuclear and cytoplasmic surveillance systems.[99,112,113] These activities not only eliminate the mistakes of RNA processing but also act with the RNA processing machineries to down-regulate gene expression through auto-regulatory loops or, by more broadly based regulatory schemes.[114] One example where the splicing apparatus is co-opted in this way is the regulated unproductive splicing and translation or RUST mechanism of mammals.[115] Here alternative splice site selection directs the inclusion of an exon containing a premature translational termination codon (PTC) that is recognized during the first round of translation to activate the NMD pathway and stimulate RNA turnover.[116,117] The magnitude of RUST regulation in the mammalian transcriptome is controversial yet several unambiguous examples clearly establish RUST as a regulatory pathway (e.g., managing the abundance of SR protein alternative splicing factors).[118,119]

A second instance where gene expression is regulated by the directed decay of unprocessed RNA was recently discovered in yeast by the Guthrie and Chanfreau laboratories. Here amino acid starvation induced by the addition of the anti-metabolite 3-aminotriazole is shown to inhibit the splicing[120] with the unspliced pre-mRNA exported to the cytoplasm and degraded by NMD.[121] Similar to RUST regulation, this nutrient regulation is pathway-specific with transcripts contributing to ribosome biogenesis being acutely sensitive and most other pre-mRNAs unaffected. As such, this uncoupling of splicing from RNA export defines a novel homeostatic response to adjust for decreased ribosome demand when translation is limited by amino acid availability. Ribosome biogenesis is controlled at multiple transcriptional and post-transcriptional levels and regulated by TOR, PKA and PKC signaling (reviewed in ref. 122). The specific response to amino acid starvation has been well studied[123,124] although the signaling molecules and targets that mediate splicing inhibition and the nuclear release of unprocessed pre-mRNA remain unknown.

DOES BBP HAVE A CYTOPLASMIC FUNCTION?

BBP binds Smy2,[23,125,126] a protein that is enriched in cytoplasmic P-bodies, sites of mRNA storage and RNA decay.[127] The BBP-Smy2 interaction is intriguing from the perspective that should pre-mRNA exit the nucleus bound with BBP, interaction with Smy2 might help segregate and turnover this intron-bearing RNA. Smy2 and a highly related second protein called Syh1 are members of the GYF family of polyproline binding proteins.[128] Peptide binding and protein purification studies show that Smy2 binds six peptides defined by the consensus PPG (F/I/L/M/V) in the C-terminus of BBP, a region much expanded in SF1[90] (see Fig. 1A,B). Smy2 was also found to bind peptides in the Prp8 splicing factor and in Eap1, a protein implicated in TOR signaling and suggested to act as a translational inhibitor.[129]

The first hint of *SMY2* function came with its identification as a dosage-dependent suppressor of a mutant Type V myosin motor protein, *myo2-66*. Type V myosin promotes intracellular sorting and asymmetry, including mRNA localization. A large number of RNAs associate with Myo2 in cytoplasmic foci that appear to be P-bodies.[130] Myo4, a second Type V myosin, directs the transport of yeast *ASH1* and other bud-localized mRNAs to the site of daughter cell synthesis.[131] Translation of Myo4-asociated mRNAs is arrested during transport. Another KH-domain RNA-binding protein, Hek2/Khd1, is present with Smy2 in Myo2 and Myo4 complexes and is implicated with Eap1 in translational arrest.[132-134] The interaction of Smy2 with a Type V myosin motor, its enrichment in P-bodies and its association with Kdh1, Eap1 and other proteins acting in translational regulation or RNA decay (i.e., Asc1, Ccr4, Kem1, Mot2, Pat1, Pop2 and Scp160 and)[126,134] strongly implicate Smy2 in the transport, sequestration and turnover of RNA.

It is tempting to speculate that Smy2 recruits BBP-bound pre-mRNA exported to the cytoplasm (by error in RNA processing or in response to specific signals) and directs this RNA via a myosin motor to the P-body. Such a scheme necessitates a cytoplasmic phase for BBP although under standard growth conditions BBP appears largely nuclear.[135] There has been no direct test for nucleocytoplasmic shuttling by BBP although GFP-localization studies suggest that its binding partner, Mud2, has a cytoplasmic phase.[135] Consistent with at least transient residence in the cytoplasm, BBP was reported as a possible ribosome-associated protein in a recent survey for regulators of translation.[136]

Additional correlative evidence suggests that Smy2 association may promote the translational arrest of BBP-bound RNA in yeast. When a deletion of the *SMY2* gene is combined with a *mud2* mutation, the expression of a reporter that requires translation through an in-frame intron significantly increases.[10] Translation is a cytoplasmic event and this intron- reporter system was initially designed to identify mutations that allow for the nuclear export of unprocessed pre-mRNA.[101] The *smy2* mutation does not alter splicing efficiency in this background thus ruling out the trivial explanation that a splicing block causes pre-mRNA levels to rise. While other explanations are possible, the data are consistent with enhanced cytoplasmic pre-mRNA stability or increased translation after *smy2* deletion when BBP binding to RNA is compromised by removal of Mud2. This experiment does not directly address the question of whether BBP is associated with the reporter pre-mRNA in the cytoplasm. However, deletion of one of two yeast genes that encode the eIF-5A translation elongation factor has been shown to improve the growth of yeast limited for BBP.[98] The molecular basis for this interaction is not known yet the observation generally supports a model in which BBP restricts the synthesis of toxic proteins and that reduced translational efficiency can partially compensate for lowered BBP activity.

DOES BBP REGULATE THE FATE OF INTRONLESS RNA?

BBP is a sequence-specific RNA binding protein that in principle binds any RNA with the UACUAAC motif. At least 375 yeast genes have one or more perfect matches to this motif within protein-coding sequence. The number of potential BBP targets increases substantially if the 5' and 3' UTR regions of protein coding genes or the transcribed portions of noncoding RNAs are considered. The protein coding capacity of the UACUAAC sequence is relevant to protein structure and may well be conserved for this reason in specific genes. However, the coding capacity of UACUAAC has no obvious bearing on the ability of BBP to bind the cognate RNA. Indeed, a number of intronless mRNAs containing the UACUAAC sequence have already been shown to copurify with BBP from yeast.[137] The biological relevance of this association has not been addressed but, based on the established or predicted BBP function, one would speculate that bound mRNAs might be preferentially retained in the nucleus or translationally repressed in the cytoplasm. The intronless mRNAs recovered with BBP function in a number of metabolic and gene expression processes but do not cluster in a way to suggest clear patterns of coordinated regulation. Nevertheless, if this indicates a true intracellular association, one anticipates that BBP removal will be necessary for efficient protein expression. Sub2 is a likely candidate for promoting BBP dissociation and in this light the suppression of a lethal *sub2* mutant by weak *msl5* (or *mud2* knockout) mutations conceivably reflects not only BBP displacement from pre-mRNA but also from intronless RNAs bearing the UACUAAC motif.

CONCLUSION

The accumulated evidence suggests that BBP/SF1 functions from the earliest steps of co-transcriptional spliceosome assembly through to the turnover of aberrant or otherwise unneeded RNA (Fig. 4). Other STAR/GSG proteins show similar multidimensional contributions in nucleic acid metabolism and function. For instance, the mammalian quaking (QKI) and Sam68 proteins are implicated in alternative RNA splicing, RNA export from the nucleus and translational regulation,[138,139] the *Drosophila* HOW protein in pre-mRNA splicing and RNA stability,[140] the *C. elegans* GLD-1 protein in translational repression and RNA stability and the *C. elegans* ASD-2 protein in alternative splicing and possibly RNA stability.[141] These proteins promote a wide range of signal mediated developmental and

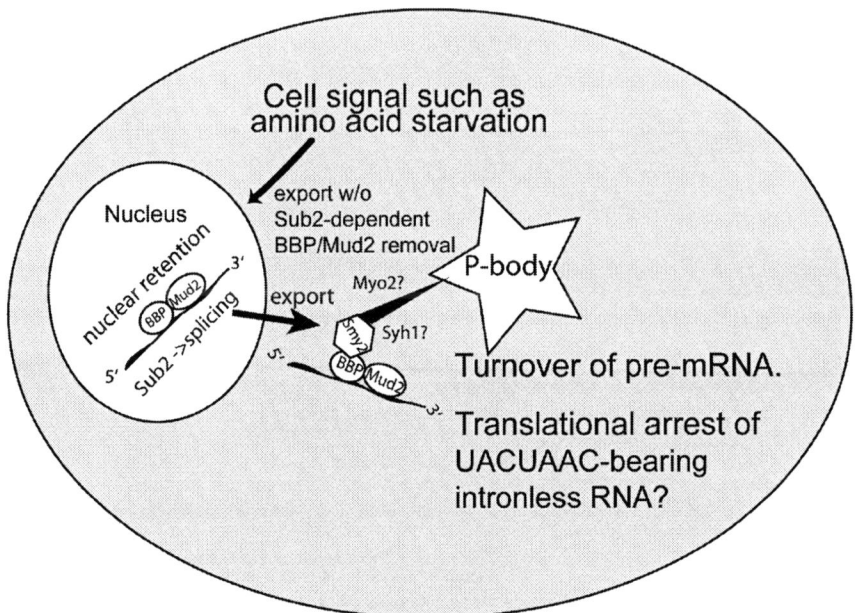

Figure 4. BBP contribution to gene expression. The accumulated evidence suggests that BBP may have multiple intracellular functions including the nuclear retention of UACUAAC-bearing RNA, facilitation of nuclear spliceosome assembly and the translational arrest or turnover of RNA in the cytoplasm. Export to the cytoplasm is presented as Sub2 dependent and the cytoplasmic P-body association directed through Smy2 and associated proteins (Syh1, Myo2).

tissue-specific events (see refs. 139,142-149 and references within). The overlap of RNA binding specificity of SF1, GLD-1, HOW, Sam68 and QKI is remarkable and has led to the suggestion that competitive binding at the pre-mRNA branchpoint may explain the influence of these proteins on splice site selection.[150] Other RNA-binding proteins, such as U2AF65/Mud2, muscleblind, SF2/ASF, SC35 and additional SR protein members also clearly contribute to the functional integration of transcription, RNA processing, RNA export, translation and RNA decay.[151-157] Much work lies ahead in defining the intricacies of RNA-protein and protein-protein interactions linking these steps of gene expression.

A number of preliminary studies with BBP raise intriguing questions that deserve further study. For instance, is BBP essential because of its role in spliceosome assembly or some other aspect of cell biochemistry? Is the BBP interaction with the CDT-binding protein Prp40 important for co-transcriptional splicing? Does amino acid starvation promote BBP-Mud2 dissociation from pre-mRNA to stimulate nuclear export or is the pre-mRNA exported to the cytoplasm bound with this heterodimer? What signaling pathways are involved in this amino-acid starvation response of splicing inhibition? Is BBP post-translationally modified similar to SF1 and if so does this modification influence its activity in the nuclear retention of pre-mRNA, splicing or other aspects of RNA fate? If cytoplasmic pre-mRNA is BBP-bound, does Smy2 (or Syh1) interaction promote translational arrest or RNA turnover? Does BBP influence expression of proteins from intronless RNA by nuclear retention of the mRNA or by translational inhibition and, if so, is Sub2 involved in this regulation? Many of the same questions can be asked of SF1 and this topic is further enriched by the presence of multiple STAR/GSG proteins with overlapping specificities in mammals.

REFERENCES

1. Wahl MC, Will CL, Luhrmann R. The spliceosome: design principles of a dynamic RNP machine. Cell 2009; 136(4):701-718.
2. Toda T, Iida A, Miwa T et al. Isolation and characterization of a novel gene encoding nuclear protein at a locus (D11S636) tightly linked to multiple endocrine neoplasia type 1 (MEN1). Hum Mol Genet 1994; 3(3):465-470.
3. Kramer A. Purification of splicing factor SF1, a heat-stable protein that functions in the assembly of a presplicing complex. Mol Cell Biol 1992; 12(10):4545-4552.
4. Rain JC, Rafi Z, Rhani Z et al. Conservation of functional domains involved in RNA binding and protein-protein interactions in human and Saccharomyces cerevisiae pre-mRNA splicing factor SF1. RNA 1998; 4(5):551-565.
5. Shitashige M, Satow R, Honda K et al. Increased susceptibility of Sf1(+/-) mice to azoxymethane-induced colon tumorigenesis. Cancer Sci 2007; 98(12):1862-1867.
6. Lukong KE, Chang KW, Khandjian EW et al. RNA-binding proteins in human genetic disease. Trends Genet 2008; 24(8):416-425.
7. Abovich N, Rosbash M. Cross-intron bridging interactions in the yeast commitment complex are conserved in mammals. Cell 1997; 89(3):403-412.
8. Zamore PD, Patton JG, Green MR. Cloning and domain structure of the mammalian splicing factor U2AF. Nature 1992; 355(6361):609-614.
9. Rutz B, Seraphin B. A dual role for BBP/ScSF1 in nuclear pre-mRNA retention and splicing. EBMO J 2000; 19(8):1873-1886.
10. Wang Q, Zhang L, Lynn B et al. A BBP-Mud2p heterodimer mediates branchpoint recognition and influences splicing substrate abundance in budding yeast. Nucleic Acids Res 2008; 36(8):2787-2798.
11. Seraphin B, Rosbash M. Identification of functional U1 snRNA-pre-mRNA complexes committed to spliceosome assembly and splicing. Cell 1989; 59(2):349-358.
12. Michaud S, Reed R. A functional association between the 5' and 3' splice site is established in the earliest prespliceosome complex (E) in mammals. Genes Dev 1993; 7(6):1008-1020.

13. Michaud S, Reed R. An ATP-independent complex commits pre-mRNA to the mammalian spliceosome assembly pathway. Genes Dev 1991; 5(12B):2534-2546.
14. Berglund JA, Fleming ML, Rosbash M. The KH domain of the branchpoint sequence binding protein determines specificity for the pre-mRNA branchpoint sequence. RNA 1998; 4(8):998-1006.
15. Garrey SM, Voelker R, Berglund JA. An extended RNA binding site for the yeast branch point-binding protein and the role of its zinc knuckle domains in RNA binding. J Biol Chem 2006; 281(37):27443-27453.
16. Buratti E, Baralle FE. Influence of RNA secondary structure on the pre-mRNA splicing process. Mol Cell Biol 2004; 24(24):10505-10514.
17. Grishin NV. KH domain: one motif, two folds. Nucleic Acids Res 2001; 29(3):638-643.
18. Liu Z, Luyten I, Bottomley MJ et al. Structural basis for recognition of the intron branch site RNA by splicing factor 1. Science 2001; 294(5544):1098-1102.
19. Garrey SM, Cass DM, Wandler AM et al. Transposition of two amino acids changes a promiscuous RNA binding protein into a sequence-specific RNA binding protein. RNA 2008; 14(1):78-88.
20. Berglund JA, Chua K, Abovich N et al. The splicing factor BBP interacts specifically with the pre-mRNA branchpoint sequence UACUAAC. Cell 1997; 89(5):781-787.
21. Keller EB, Noon WA. Intron splicing: a conserved internal signal in introns of animal pre-mRNAs. Proc Natl Acad Sci USA 1984; 81(23):7417-7420.
22. Wang Z, Burge CB. Splicing regulation: from a parts list of regulatory elements to an integrated splicing code. RNA 2008; 14(5):802-813.
23. Fromont-Racine M, Rain JC, Legrain P. Toward a functional analysis of the yeast genome through exhaustive two- hybrid screens [see comments]. Nat Genet 1997; 16(3):277-282.
24. Berglund JA, Abovich N, Rosbash M. A cooperative interaction between U2AF65 and mBBP/SF1 facilitates branchpoint region recognition. Genes Dev 1998; 12(6):858-867.
25. Abovich N, Liao XC, Rosbash M. The yeast MUD2 protein: an interaction with PRP11 defines a bridge between commitment complexes and U2 snRNP addition. Genes Dev 1994; 8(7):843-854.
26. Clark TA, Sugnet CW, Ares M Jr. Genomewide analysis of mRNA processing in yeast using splicing-specific microarrays. Science 2002; 296(5569):907-910.
27. Kielkopf CL, Lucke S, Green MR. U2AF homology motifs: protein recognition in the RRM world. Genes Dev 2004; 18(13):1513-1526.
28. Jenkins JL, Shen H, Green MR et al. Solution conformation and thermodynamic characteristics of RNA binding by the splicing factor U2AF65. J Biol Chem 2008; 283(48):33641-33649.
29. Gaur RK, Valcarcel J, Green MR. Sequential recognition of the pre-mRNA branch point by U2AF65 and a novel spliceosome-associated 28-kDa protein. RNA 1995; 1(4):407-417.
30. Banerjee H, Rahn A, Davis W et al. Sex lethal and U2 small nuclear ribonucleoprotein auxiliary factor (U2AF65) recognize polypyrimidine tracts using multiple modes of binding. RNA 2003; 9(1):88-99.
31. Valcarcel J, Gaur RK, Singh R et al. Interaction of U2AF65 RS region with pre-mRNA branch point and promotion of base pairing with U2 snRNA [corrected] [published erratum appears in Science 1996; 274(5284):21]. Science 1996; 273(5282):1706-1709.
32. Selenko P, Gregorovic G, Sprangers R et al. Structural basis for the molecular recognition between human splicing factors U2AF65 and SF1/mBBP. Mol Cell 2003; 11(4):965-976.
33. Wang X, Bruderer S, Rafi Z et al. Phosphorylation of splicing factor SF1 on Ser20 by cGMP-dependent protein kinase regulates spliceosome assembly. EBMO J 1999; 18(16):4549-4559.
34. Corsini L, Bonnal S, Basquin J et al. U2AF-homology motif interactions are required for alternative splicing regulation by SPF45. Nat Struct Mol Biol 2007; 14(7):620-629.
35. Rain JC, Legrain P. In vivo commitment to splicing in yeast involves the nucleotide upstream from the branch site conserved sequence and the Mud2 protein. EBMO J 1997; 16(7):1759-1771.
36. Rymond BC, Torrey DD, Rosbash M. A novel role for the 3' region of introns in pre-mRNA splicing of Saccharomyces cerevisiae. Genes Dev 1987; 1(3):238-246.
37. Rutz B, Seraphin B. Transient interaction of BBP/ScSF1 and Mud2 with the splicing machinery affects the kinetics of spliceosome assembly. RNA 1999; 5(6):819-831.
38. Zamore PD, Green MR. Identification, purification and biochemical characterization of U2 small nuclear ribonucleoprotein auxiliary factor. Proc Natl Acad Sci USA 1989; 86(23):9243-9247.
39. Wu S, Romfo CM, Nilsen TW et al. Functional recognition of the 3' splice site AG by the splicing factor U2AF35. Nature 1999; 402(6763):832-835.
40. Zorio DA, Blumenthal T. Both subunits of U2AF recognize the 3' splice site in Caenorhabditis elegans. Nature 1999; 402(6763):835-838.
41. Merendino L, Guth S, Bilbao D et al. Inhibition of msl-2 splicing by Sex-lethal reveals interaction between U2AF35 and the 3' splice site AG. Nature 1999; 402(6763):838-841.
42. Huang T, Vilardell J, Query CC. Prespliceosome formation in S.pombe requires a stable complex of SF1-U2AF(59)-U2AF(23). EBMO J 2002; 21(20):5516-5526.

43. Rino J, Desterro JM, Pacheco TR et al. Splicing factors SF1 and U2AF associate in extraspliceosomal complexes. Mol Cell Biol 2008; 28(9):3045-3057.
44. Linder P. Dead-box proteins: a family affair—active and passive players in RNP-remodeling. Nucleic Acids Res 2006; 34(15):4168-4180.
45. Smith DJ, Query CC, Konarska MM. "Nought may endure but mutability": spliceosome dynamics and the regulation of splicing. Mol Cell 2008; 30(6):657-666.
46. Puig O, Gottschalk A, Fabrizio P et al. Interaction of the U1 snRNP with nonconserved intronic sequences affects 5' splice site selection. Genes Dev 1999; 13(5):569-580.
47. Zhang D, Rosbash M. Identification of eight proteins that cross-link to pre-mRNA in the yeast commitment complex. Genes Dev 1999; 13(5):581-592.
48. Colot HV, Stutz F, Rosbash M. The yeast splicing factor Mud13p is a commitment complex component and corresponds to CBP20, the small subunit of the nuclear cap-binding complex. Genes Dev 1996; 10(13):1699-1708.
49. Lewis JD, Gorlich D, Mattaj IW. A yeast cap binding protein complex (yCBC) acts at an early step in pre mRNA splicing. Nucleic Acids Res 1996; 24(17):3332-3336.
50. Hage R, Tung L, Du H et al. A targeted bypass screen identifies Ynl187p, Prp42p, Snu71p and Cbp80p for stable U1 snRNP/Pre-mRNA interaction. Mol Cell Biol 2009; 29(14):3941-3952.
51. Du H, Rosbash M. The U1 snRNP protein U1C recognizes the 5' splice site in the absence of base pairing. Nature 2002; 419(6902):86-90.
52. Balzer RJ, Henry MF. Snu56p is required for Mer1p-activated meiotic splicing. Mol Cell Biol 2008; 28(8):2497-2508.
53. Corsini L, Hothorn M, Stier G et al. Dimerization and protein binding specificity of the U2AF homology motif of the splicing factor Puf60. J Biol Chem 2009; 284(1):630-639.
54. Wang Q, Rymond BC. Rds3p is required for stable U2 snRNP recruitment to the splicing apparatus. Molecular and Cellular Biology 2003; 23(20):7339-7349.
55. Wang Q, He J, Lynn B et al. Interactions of the Yeast SF3b Splicing Factor. Mol Cell Biol 2005; 25(24):10745-10754.
56. Dziembowski A, Ventura AP, Rutz B et al. Proteomic analysis identifies a new complex required for nuclear pre-mRNA retention and splicing. EBMO J 2004; 23(24):4847-4856.
57. Gozani O, Potashkin J, Reed R. A potential role for U2AF-SAP 155 interactions in recruiting U2 snRNP to the branch site. Mol Cell Biol 1998; 18(8):4752-4760.
58. Spadaccini R, Reidt U, Dybkov O et al. Biochemical and NMR analyses of an SF3b155-p14-U2AF-RNA interaction network involved in branch point definition during pre-mRNA splicing. RNA 2006; 12(3):410-425.
59. Hastings ML, Allemand E, Duelli DM et al. Control of pre-mRNA splicing by the general splicing factors PUF60 and U2AF65. PLoS ONE 2007; 2(6):e538.
60. Das BK, Xia L, Palandjian L et al. Characterization of a protein complex containing spliceosomal proteins SAPs 49, 130, 145 and 155. Mol Cell Biol 1999; 19(10):6796-6802.
61. Gozani O, Feld R, Reed R. Evidence that sequence-independent binding of highly conserved U2 snRNP proteins upstream of the branch site is required for assembly of spliceosomal complex A. Genes Dev 1996; 10(2):233-243.
62. Xu YZ, Newnham CM, Kameoka S et al. Prp5 bridges U1 and U2 snRNPs and enables stable U2 snRNP association with intron RNA. EBMO J 2004; 23(2):376-385.
63. Perriman R, Barta I, Voeltz GK et al. ATP requirement for Prp5p function is determined by Cus2p and the structure of U2 small nuclear RNA. Proc Natl Acad Sci USA 2003; 100(24):13857-13862.
64. Perriman RJ, Ares M Jr. Rearrangement of competing U2 RNA helices within the spliceosome promotes multiple steps in splicing. Genes Dev 2007; 21(7):811-820.
65. Xu YZ, Query CC. Competition between the ATPase Prp5 and branch region-U2 snRNA pairing modulates the fidelity of spliceosome assembly. Mol Cell 2007; 28(5):838-849.
66. Strasser K, Masuda S, Mason P et al. TREX is a conserved complex coupling transcription with messenger RNA export. Nature 2002; 417(6886):304-308.
67. Shen H, Zheng X, Shen J et al. Distinct activities of the DExD/H-box splicing factor hUAP56 facilitate stepwise assembly of the spliceosome. Genes Dev 2008; 22(13):1796-1803.
68. Kelly SM, Corbett AH. Messenger RNA export from the nucleus: a series of molecular wardrobe changes. Traffic 2009; 10(9):1199-1208.
69. Abruzzi KC, Lacadie S, Rosbash M. Biochemical analysis of TREX complex recruitment to intronless and intron-containing yeast genes. EBMO J 2004; 23(13):2620-2631.
70. Kistler AL, Guthrie C. Deletion of MUD2, the yeast homolog of U2AF65, can bypass the requirement for sub2, an essential spliceosomal ATPase. Genes Dev 2001; 15(1):42-49.
71. Libri D, Graziani N, Saguez C et al. Multiple roles for the yeast SUB2/yUAP56 gene in splicing. Genes Dev 2001; 15(1):36-41.

72. Pandit S, Wang D, Fu XD. Functional integration of transcriptional and RNA processing machineries. Curr Opin Cell Biol 2008; 20(3):260-265.
73. Moore MJ, Proudfoot NJ. Pre-mRNA processing reaches back to transcription and ahead to translation. Cell 2009; 136(4):688-700.
74. Tardiff DF, Rosbash M. Arrested yeast splicing complexes indicate stepwise snRNP recruitment during in vivo spliceosome assembly. RNA 2006; 12(6):968-979.
75. Pandya-Jones A, Black DL. Co-transcriptional splicing of constitutive and alternative exons. RNA 2009; 15(10):1896-1908.
76. Gornemann J, Kotovic KM, Hujer K et al. Co-transcriptional spliceosome assembly occurs in a stepwise fashion and requires the cap binding complex. Mol Cell 2005; 19(1):53-63.
77. Kotovic KM, Lockshon D, Boric L et al. Co-transcriptional recruitment of the U1 snRNP to intron-containing genes in yeast. Mol Cell Biol 2003; 23(16):5768-5779.
78. Lacadie SA, Rosbash M. Co-transcriptional spliceosome assembly dynamics and the role of U1 snRNA:5'ss base pairing in yeast. Mol Cell 2005; 19(1):65-75.
79. Lacadie SA, Tardiff DF, Kadener S et al. In vivo commitment to yeast co-transcriptional splicing is sensitive to transcription elongation mutants. Genes Dev 2006; 20(15):2055-2066.
80. Moore MJ, Schwartzfarb EM, Silver PA et al. Differential recruitment of the splicing machinery during transcription predicts genome-wide patterns of mRNA splicing. Mol Cell 2006; 24(6):903-915.
81. Swinburne IA, Meyer CA, Liu XS et al. Genomic localization of RNA binding proteins reveals links between pre-mRNA processing and transcription. Genome Res 2006; 16(7):912-921.
82. Listerman I, Sapra AK, Neugebauer KM. Co-transcriptional coupling of splicing factor recruitment and precursor messenger RNA splicing in mammalian cells. Nat Struct Mol Biol 2006; 13(9):815-822.
83. Sapra AK, Anko ML, Grishina I et al. SR protein family members display diverse activities in the formation of nascent and mature mRNPs in vivo. Mol Cell 2009; 34(2):179-190.
84. Stevens SW, Ryan DE, Ge HY et al. Composition and functional characterization of the yeast spliceosomal penta-snRNP. Mol Cell 2002; 9(1):31-44.
85. Behzadnia N, Hartmuth K, Will CL et al. Functional spliceosomal A complexes can be assembled in vitro in the absence of a penta-snRNP. RNA 2006; 12(9):1738-1746.
86. Greenleaf AL. Positive patches and negative noodles: linking RNA processing to transcription? Trends Biochem Sci 1993; 18(4):117-119.
87. Natalizio BJ, Robson-Dixon ND, Garcia-Blanco MA. The Carboxyl-terminal Domain of RNA Polymerase II Is Not Sufficient to Enhance the Efficiency of Pre-mRNA Capping or Splicing in the Context of a Different Polymerase. J Biol Chem 2009; 284(13):8692-8702.
88. Bushnell DA, Cramer P, Kornberg RD. Structural basis of transcription: alpha-amanitin-RNA polymerase II cocrystal at 2.8 A resolution. Proc Natl Acad Sci USA 2002; 99(3):1218-1222.
89. Morris DP, Greenleaf AL. The splicing factor, Prp40, binds the phosphorylated carboxyl-terminal domain of RNA polymerase II. J Biol Chem 2000; 275(51):39935-39943.
90. Lin KT, Lu RM, Tarn WY. The WW domain-containing proteins interact with the early spliceosome and participate in pre-mRNA splicing in vivo. Mol Cell Biol 2004; 24(20):9176-9185.
91. Grainger RJ, Beggs JD. Prp8 protein: at the heart of the spliceosome. RNA 2005; 11(5):533-557.
92. Jobert L, Pinzon N, Van Herreweghe E et al. Human U1 snRNA forms a new chromatin-associated snRNP with TAF15. EMBO Rep 2009; 10(5):494-500.
93. de la Mata M, Alonso CR, Kadener S et al. A slow RNA polymerase II affects alternative splicing in vivo. Mol Cell 2003; 12(2):525-532.
94. Howe KJ, Kane CM, Ares M Jr. Perturbation of transcription elongation influences the fidelity of internal exon inclusion in Saccharomyces cerevisiae. RNA 2003; 9(8):993-1006.
95. Kornblihtt AR, Schor IE, Allo M et al. When chromatin meets splicing. Nat Struct Mol Biol 2009; 16(9):902-903.
96. Pleiss JA, Whitworth GB, Bergkessel M et al. Transcript specificity in yeast pre-mRNA splicing revealed by mutations in core spliceosomal components. PLoS Biol 2007; 5(4):e90.
97. Tanackovic G, Kramer A. Human splicing factor SF3a, but not SF1, is essential for pre-mRNA splicing in vivo. Mol Biol Cell 2005; 16(3):1366-1377.
98. Wilmes GM, Bergkessel M, Bandyopadhyay S et al. A genetic interaction map of RNA-processing factors reveals links between Sem1/Dss1-containing complexes and mRNA export and splicing. Mol Cell 2008; 32(5):735-746.
99. Houseley J, LaCava J, Tollervey D. RNA-quality control by the exosome. Nat Rev Mol Cell Biol 2006; 7(7):529-539.
100. Tardiff DF, Lacadie SA, Rosbash M. A genome-wide analysis indicates that yeast pre-mRNA splicing is predominantly post-transcriptional. Mol Cell 2006; 24(6):917-929.
101. Legrain P, Rosbash M. Some cis- and trans-acting mutants for splicing target pre-mRNA to the cytoplasm. Cell 1989; 57(4):573-583.

102. Abovich N, Legrain P, Rosbash M. The yeast PRP6 gene encodes a U4/U6 small nuclear ribonucleoprotein particle (snRNP) protein and the PRP9 gene encodes a protein required for U2 snRNP binding. Mol Cell Biol 1990; 10(12):6417-6425.
103. Scherrer FW Jr, Spingola M. A subset of Mer1p-dependent introns requires Bud13p for splicing activation and nuclear retention. RNA 2006; 12(7):1361-1372.
104. Spingola M, Armisen J, Ares M Jr. Mer1p is a modular splicing factor whose function depends on the conserved U2 snRNP protein Snu17p. Nucleic Acids Res 2004; 32(3):1242-1250.
105. Will CL, Schneider C, MacMillan AM et al. A novel U2 and U11/U12 snRNP protein that associates with the pre-mRNA branch site. EBMO J 2001; 20(16):4536-4546.
106. Gottschalk A, Bartels C, Neubauer G et al. A novel yeast U2 snRNP protein, Snu17p, is required for the first catalytic step of splicing and for progression of spliceosome assembly. Mol Cell Biol 2001; 21(9):3037-3046.
107. Kotake Y, Sagane K, Owa T et al. Splicing factor SF3b as a target of antitumor natural product pladienolide. Nature Chemical Genetics 2007; 3.
108. Kaida D, Motoyoshi H, Tashiro E et al. Spliceostatin A targets SF3b and inhibits both splicing and nuclear retention of pre-mRNA. Nature Chemical Genetics 2007; 3.
109. Palancade B, Zuccolo M, Loeillet S et al. Pml39, a novel protein of the nuclear periphery required for nuclear retention of improper messenger ribonucleoparticles. Mol Biol Cell 2005; 16(11):5258-5268.
110. Casolari JM, Silver PA. Guardian at the gate: preventing unspliced pre-mRNA export. Trends in Cell Biology 2004; 14(5):222-225.
111. Galy V, Gadal O, Fromont-Racine M et al. Nuclear retention of unspliced mRNAs in yeast is mediated by perinuclear Mlp1. Cell 2004; 116(1):63-73.
112. Sommer P, Nehrbass U. Quality control of messenger ribonucleoprotein particles in the nucleus and at the pore. Curr Opin Cell Biol 2005; 17(3):294-301.
113. Bruno I, Wilkinson MF. P-bodies react to stress and nonsense. Cell 2006; 125(6):1036-1038.
114. McKee AE, Silver PA. Systems perspectives on mRNA processing. Cell Res 2007; 17(7):581-590.
115. Lareau LF, Brooks AN, Soergel DA et al. The coupling of alternative splicing and nonsense-mediated mRNA decay. Adv Exp Med Biol 2007; 623:190-211.
116. Rebbapragada I, Lykke-Andersen J. Execution of nonsense-mediated mRNA decay: what defines a substrate? Curr Opin Cell Biol 2009; 21(3):394-402.
117. Chang YF, Imam JS, Wilkinson MF. The nonsense-mediated decay RNA surveillance pathway. Annu Rev Biochem 2007; 76:51-74.
118. McGlincy NJ, Smith CW. Alternative splicing resulting in nonsense-mediated mRNA decay: what is the meaning of nonsense? Trends Biochem Sci 2008; 33(8):385-393.
119. Ni JZ, Grate L, Donohue JP et al. Ultraconserved elements are associated with homeostatic control of splicing regulators by alternative splicing and nonsense-mediated decay. Genes Dev 2007; 21(6):708-718.
120. Pleiss JA, Whitworth GB, Bergkessel M et al. Rapid, transcript-specific changes in splicing in response to environmental stress. Mol Cell 2007; 27(6):928-937.
121. Sayani S, Janis M, Lee CY et al. Widespread impact of nonsense-mediated mRNA decay on the yeast intronome. Mol Cell 2008; 31(3):360-370.
122. Henras AK, Soudet J, Gerus M et al. The post-transcriptional steps of eukaryotic ribosome biogenesis. Cell Mol Life Sci 2008; 65(15):2334-2359.
123. Halbeisen RE, Gerber AP. Stress-Dependent Coordination of Transcriptome and Translatome in Yeast. PLoS Biol 2009; 7(5):e105.
124. Zaman S, Lippman SI, Zhao X et al. How Saccharomyces responds to nutrients. Annu Rev Genet 2008; 42:27-81.
125. Kofler M, Motzny K, Freund C. GYF domain proteomics reveals interaction sites in known and novel target proteins. Mol Cell Proteomics 2005; 4(11):1797-1811.
126. Georgiev A, Sjostrom M, Wieslander A. Binding specificities of the GYF domains from two Saccharomyces cerevisiae Paralogs. Protein Eng Des Sel 2007.
127. Parker R, Sheth U. P bodies and the control of mRNA translation and degradation. Mol Cell 2007; 25(5):635-646.
128. Kofler MM, Freund C. The GYF domain. Febs J 2006; 273(2):245-256.
129. Cosentino GP, Schmelzle T, Haghighat A et al. Eap1p, a novel eukaryotic translation initiation factor 4E-associated protein in Saccharomyces cerevisiae. Mol Cell Biol 2000; 20(13):4604-4613.
130. Chang W, Zaarour RF, Reck-Peterson S et al. Myo2p, a class V myosin in budding yeast, associates with a large ribonucleic acid-protein complex that contains mRNAs and subunits of the RNA-processing body. RNA 2008; 14(3):491-502.
131. Paquin N, Chartrand P. Local regulation of mRNA translation: new insights from the bud. Trends Cell Biol 2008; 18(3):105-111.

132. Paquin N, Menade M, Poirier G et al. Local activation of yeast ASH1 mRNA translation through phosphorylation of Khd1p by the casein kinase Yck1p. Mol Cell 2007; 26(6):795-809.
133. Hasegawa Y, Irie K, Gerber AP. Distinct roles for Khd1p in the localization and expression of bud-localized mRNAs in yeast. RNA 2008; 14(11):2333-2347.
134. Sezen B, Seedorf M, Schiebel E. The SESA network links duplication of the yeast centrosome with the protein translation machinery. Genes Dev 2009; 23(13):1559-1570.
135. Huh WK, Falvo JV, Gerke LC et al. Global analysis of protein localization in budding yeast. Nature 2003; 425(6959):686-691.
136. Fleischer TC, Weaver CM, McAfee KJ et al. Systematic identification and functional screens of uncharacterized proteins associated with eukaryotic ribosomal complexes. Genes Dev 2006; 20(10):1294-1307.
137. Hogan DJ, Riordan DP, Gerber AP et al. Diverse RNA-binding proteins interact with functionally related sets of RNAs, suggesting an extensive regulatory system. PLoS Biol 2008; 6(10):e255.
138. Chenard CA, Richard S. New implications for the QUAKING RNA binding protein in human disease. J Neurosci Res 2008; 86(2):233-242.
139. Sette C, Messina V, Paronetto MP. Sam68: A New STAR in the Male Fertility Firmament. J Androl 2009.
140. Volk T, Israeli D, Nir R et al. Tissue development and RNA control: "HOW" is it coordinated? Trends Genet 2008; 24(2):94-101.
141. Ohno G, Hagiwara M, Kuroyanagi H. STAR family RNA-binding protein ASD-2 regulates developmental switching of mutually exclusive alternative splicing in vivo. Genes Dev 2008; 22(3):360-374.
142. Stoss O, Olbrich M, Hartmann AM et al. The STAR/GSG family protein rSLM-2 regulates the selection of alternative splice sites. J Biol Chem 2001; 276(12):8665-8673.
143. Chawla G, Lin CH, Han A et al. Sam68 regulates a set of alternatively spliced exons during neurogenesis. Mol Cell Biol 2009; 29(1):201-213.
144. Paronetto MP, Achsel T, Massiello A et al. The RNA-binding protein Sam68 modulates the alternative splicing of Bcl-x. J Cell Biol 2007; 176(7):929-939.
145. Saccomanno L, Loushin C, Jan E et al. The STAR protein QKI-6 is a translational repressor. Proc Natl Acad Sci USA 1999; 96(22):12605-12610.
146. Lee MH, Schedl T. Translation repression by GLD-1 protects its mRNA targets from nonsense-mediated mRNA decay in C. elegans. Genes Dev 2004; 18(9):1047-1059.
147. Larocque D, Pilotte J, Chen T et al. Nuclear retention of MBP mRNAs in the quaking viable mice. Neuron 2002; 36(5):815-829.
148. Galarneau A, Richard S. The STAR RNA binding proteins GLD-1, QKI, SAM68 and SLM-2 bind bipartite RNA motifs. BMC Mol Biol 2009; 10:47.
149. Lukong KE, Richard S. Sam68, the KH domain-containing superSTAR. Biochim Biophys Acta 2003; 1653(2):73-86.
150. Ryder SP, Frater LA, Abramovitz DL et al. RNA target specificity of the STAR/GSG domain post-transcriptional regulatory protein GLD-1. Nat Struct Mol Biol 2004; 11(1):20-28.
151. Pascual M, Vicente M, Monferrer L et al. The Muscleblind family of proteins: an emerging class of regulators of developmentally programmed alternative splicing. Differentiation 2006; 74(2-3):65-80.
152. Shepard PJ, Hertel KJ. The SR protein family. Genome Biol 2009; 10(10):242.
153. Keene JD. RNA regulons: coordination of post-transcriptional events. Nat Rev Genet 2007; 8(7):533-543.
154. Gama-Carvalho M, Barbosa-Morais NL, Brodsky AS et al. Genome-wide identification of functionally distinct subsets of cellular mRNAs associated with two nucleocytoplasmic-shuttling mammalian splicing factors. Genome Biol 2006; 7(11):R113.
155. Blanchette M, Labourier E, Green RE et al. Genome-wide analysis reveals an unexpected function for the Drosophila splicing factor U2AF50 in the nuclear export of intronless mRNAs. Mol Cell 2004; 14(6):775-786.
156. Allemand E, Batsche E, Muchardt C. Splicing, transcription and chromatin: a menage a trois. Curr Opin Genet Dev 2008; 18(2):145-151.
157. Zhong XY, Wang P, Han J et al. SR proteins in vertical integration of gene expression from transcription to RNA processing to translation. Mol Cell 2009; 35(1):1-10.
158. Pandit S, Lynn B, Rymond BC. Inhibition of a spliceosome turnover pathway suppresses splicing defects. Proc Natl Acad Sci USA 2006; 103(37):13700-13705.
159. Pandit S, Paul S, Zhang L et al. Spp382p interacts with multiple yeast splicing factors, including possible regulators of Prp43 DExD/H-Box protein function. Genetics 2009; 183(1):195-206.
160. Ohi MD, Link AJ, Ren L et al. Proteomics analysis reveals stable multiprotein complexes in both fission and budding yeasts containing Myb-related Cdc5p/Cef1p, novel pre-mRNA splicing factors and snRNAs. Mol Cell Biol 2002; 22(7):2011-2024.
161. Wang Q, Hobbs K, Lynn B et al. The Clf1p splicing factor promotes spliceosome assembly through N-terminal tetratricopeptide repeat contacts. J Biol Chem 2003; 278(10):7875-7883.

CHAPTER 10

REACHING FOR THE STARS
Linking RNA Binding Proteins to Diseases

Stéphane Richard*

Abstract: The prototype STAR (Signal Transduction and Activation of RNA) protein is Sam68, the S̲rc-a̲ssociated substrate during m̲itosis of 6̲8 kDa. Sam68, like all other STAR proteins, belongs to the large class of heteronuclear ribonucleoprotein particle K (hnRNP K) homology (KH) domain family of RNA-binding proteins. The KH domain is an evolutionarily conserved RNA binding domain that consists of 70-100 amino acids. The KH domain is one of the most prevalent RNA binding domains that directly contacts single-stranded RNA with a signature topology. Sam68 contains a single KH domain that harbors additional conserved N- and C-terminal sequences also required for RNA binding specificity and dimerization. Sam68 frequently contains post-translational modifications including serine/threonine, tyrosine phosphorylation, lysine acetylation, arginine methylation and sumoylation. The phosphorylation of Sam68 or its association with SH3 domain containing proteins has been shown to influence its RNA binding activity. Hence Sam68 behaves as a STAR protein, whereby extracellular signals influence its ability to regulate RNA metabolism. Studies in mice have revealed physiological roles linking Sam68 to osteoporosis, cancer, infertility and ataxia. The role of Sam68, a closely related family member quaking (QKI), the KH domain and their links with human disease will be discussed in the present chapter.

Sam68: ITS DISCOVERY AND NOMENCLATURE

The official symbol for Sam68 in the National Center for Biotechnology Information (NCBI) database is abbreviated as KHDRBS1 for K̲H D̲omain containing, R̲NA B̲inding, S̲ignal transduction associated 1̲. Sam68 has two close mammalian members, known

*Stéphane Richard—Segal Cancer Centre, Lady Davis Institute, 3755 Côte Ste-Catherine Road, Montréal, Québec, Canada H3T 1E2. Email: stephane.richard@mcgill.ca

Post-Transcriptional Regulation by STAR Proteins: Control of RNA Metabolism in Development and Disease, edited by Talila Volk and Karen Artzt.
©2010 Landes Bioscience and Springer Science+Business Media.

as Sam68-like mammalian proteins 1 and 2 (slm1 and slm2),[1,2] now officially called KHDRBS2 and 3, respectively. The human *sam68* gene contains 9 exons and is located on chromosome 1p32 spanning 30,834 base pairs. In 1992, Sam68, then called p62, was immunopurified from v-Src-transformed NIH3T3 cells by using anti-phosphotyrosine antibodies. An amino acid sequence from the purified protein was then used to design oligonucleotides to screen a human placenta library.[3] Two years later, five different groups identified p62 as an SH3 domain binding protein and as a substrate of Src family kinases.[4-8] Courtneidge and Shalloway noticed that 'p62' migrated at 68 kDa and was a Src substrate during mitosis, hence the name Sam68.[9,10] Over the years, Sam68 has also been shown to be a substrate of other tyrosine kinases and of Src kinases during cell processes other than mitosis.[11-14] Thus Sam68 has broader roles than implied by its name. For example, Sam68 is a substrate of the BReast tumor Kinase (BRK, also called protein tyrosine kinase PTK6) during epidermal growth factor (EGF) stimulation[12] and it is a substrate of Src kinases near the plasma membrane during cell attachment.[11] Sam68 also participates in T-cell receptor,[15,16] leptin[17] and insulin receptor signaling.[18,19]

The *sam68* gene has been cloned from many species including human, rodents, chimpanzee, cow, fish, frogs, dog and chicken (Fig. 1). The human Sam68 cDNA encodes 443 amino acids[3] and the protein migrates on SDS polyacrylamide gels with a mass of 68kDa most likely because of its acidic C-terminus. Homologs for Sam68 do not exist in *C. elegans*, *Drosophila* and yeast. However, *Drosophila*, does express Sam50, a protein that bears 50% sequence identity in the STAR domain of human Sam68 and is the closest ortholog.[20]

THE KH DOMAIN

STAR proteins contain a KH domain flanked by conserved sequences ~80 and ~30 amino acids referred to as the N-terminal of KH (NK) and the C-terminal of KH (CK) regions, respectively (Fig. 2). The entire region is called the STAR domain or GSG domain or maxi-KH domain. The NK and CK regions are also called the QUA1 and QUA2 regions in the QKI protein (see Fig. 2). The role of these extended conserved domains is to mediate homodimerization and to extend the nucleic acid region of recognition.[21]

The KH domain is conserved in a variety of organisms ranging from bacteria, archaea and eukaryotes.[22,23] The KH domain has been shown to bind single-stranded RNA as well as DNA with a $\beta1\alpha1\alpha2\beta2\beta3$ topology (Type I) and $\alpha1\beta1\beta2\alpha2\alpha3\beta3$ topology (Type II). The feature of the KH domain is the conserved GXXG loop that provides close contact with the phosphate groups, such that the neighboring nucleotides can directly interact with RNA.[24] KH domains are often repeated in proteins and each domain exhibits its own RNA binding preference and affinity. As such, multiple KH domains within the same protein likely cooperate together to exhibit a multitude of RNA binding affinities to one protein. For example, the K-homology splicing regulator protein (KSRP) contains four KH domains and each domain exhibits its own specificity and affinity.[25] The presence of multiple copies of the KH domain within proteins is an indication that KH-type RNA binding proteins likely require more than one KH domain for high affinity interactions with RNA in vivo. Subsequently, KH domains are also necessary for homo-dimerization and STAR proteins require this property for their RNA binding activity.[21] The Sam68 KH domain loops 1 and 4 are necessary for this homo-oligomerization. These oligomers are disrupted by tyrosine phosphorylation by Src family tyrosine kinases.[21] The dimerization

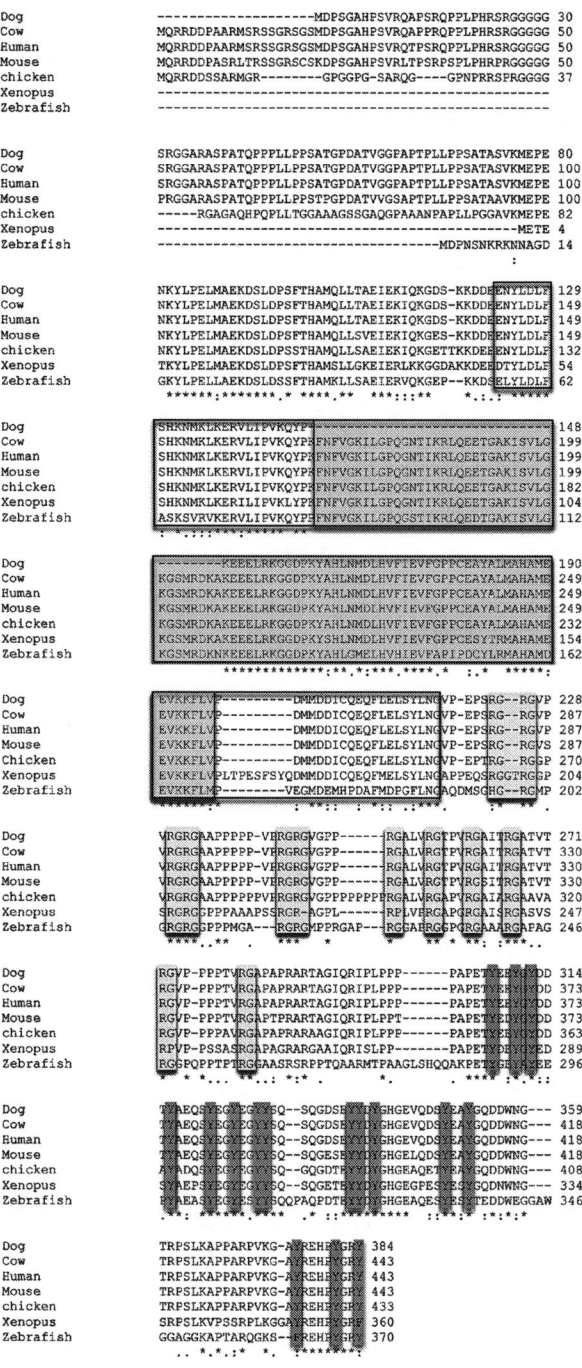

Figure 1. Multiple sequence alignment of Sam68 orthologues performed using ClustalW. The STAR domain is boxed in white, the KH domain is boxed in medium grey or blue, the arginine-glycine (RG) repeats are in light grey or yellow and the C-terminal tyrosine are in dark grey or red. A color version of this figure is available at www.landesbioscience.com/curie.

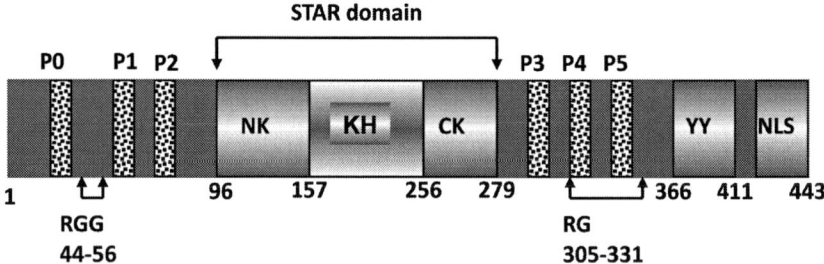

Figure 2. Schematic diagram representing the structural/functional domains of Sam68. The protein is composed of the STAR domain, a tripartite of a single RNA binding KH domain, flanked with the NK (N-terminal of KH) and the CK (C-terminal of KH) regions; six consensus proline-rich motifs (shaded regions P0-P5); RGG boxes; C-terminal tyrosine-rich domain (YY) and a nuclear localization signal (NLS). Relative amino acids positions are indicated.

sequences for QKI were mapped to the NK region which contains a predicted coiled-coil region. Interestingly, an ethylnitrosourea-induced mutation that alters glutamic acid 48 (E48G) and disrupts the coiled-coil region abolishing dimerization. This is likely the molecular defect causing embryonic lethality in these mice.[21,26]

Sam68 RNA TARGETS

Sam68 has been shown to bind poly (U) and poly (A) homopolymeric RNAs (reviewed in ref. 10). Selection of RNA aptamers using SELEX (systematic evolution of ligands by exponential enrichment) has been performed with several STAR proteins. Recombinant Sam68 was shown to associate with high affinity to RNA sequences containing UAAA or UUUA sequences.[27] SLM-2 selected a bipartite consensus sequence with direct repeats of U(U/A)AA and interestingly, Sam68 also bound this sequence.[28] The RNA targets that QKI selected were also bipartite consensus sequences with a core and half site sequence (NACUAAY-N(1-20)-UAAY, where N and Y represent any nucleotide and pyrimidine, respectively).[29] With this consensus sequence a total of 1430 mRNAs were predicted to be potential QKI RNA targets.[29]

RNA targets for STAR proteins were also identified using various other approaches. The splicing factor SF1 (also called mammalian branch point binding protein)[30] recognizes the RNA branch point sequence UACUAAC discovered by cross-linking studies.[31] Based on sequence comparison between SF1 and GLD-1, a consensus sequence was determined for GLD-1 with the UACUCA sequence.[32] Sam68 RNA targets were identified using differential display and cDNA representational difference analysis and led to the identification of 29 potential RNA-binding targets including mRNAs encoding DAP3/IRCP, nucleolar protein-p40, hnRNP A2/B1, PAP/ANX5, PBP/PEA-BP and β-actin.[33] Cross-linking protein and RNA led to the identification of 23 putative Sam68 targets.[34] The expression of 418 mRNAs was differentially expressed in testis between Sam68 deficient and wild type mice.[35] Select mRNAs such as *Klk1, Nedd1, Park2, Spag16* and *Spdya* were validated as Sam68 RNA targets during spermatogenesis.[35] A repertoire of neuronal Sam68-associated mRNAs including the elongation factor *eEF1A* was identified using co-immunoprecipitation followed by microarray analysis.[36]

Sam68 has been shown to regulate alternative splice selection by recognizing RNA sequences neighboring the included/excluded exon(s).[37] Sam68 was proposed to flag certain exons that are to be alternatively spliced.[38] It has been shown to regulate the inclusion of the variable exon 5 (V5) of CD44 correlating with cell migration potential.[37,39] Sam68 has been shown to play a role in cell survival by regulating the alternative splicing of BclxL.[13] Sam68 has also been implicated in the regulation of androgen induced alternative splicing in prostate cancer cells.[40] It has been demonstrated to regulate the leptin-induced alternative splicing of the leptin receptor mRNA.[41] Sam68 was also shown to regulate a set of alternative spliced exons during neurogenesis.[42] The absence of Sam68 modified the alternative splicing of exons across 24 different genes in Neuro2a cell line.[42] Sam68 was shown to participate in neural stem cell differentiation and proliferation. This coincided with the modulation of the alternative splicing of tenascin-C, an extracellular matrix glycoprotein.[43] The challenge that lies ahead is to link the RNA targets with the particular pathway(s) and physiological response regulated by Sam68, especially those identified using mice models.

Sam68 CELLULAR LOCALIZATION

In most cell types, Sam68 is predominantly nuclear and this is directed by a nuclear localization signal (NLS) embedded in the last 24 amino acids ([420]RPSLKAPPARPVKGAYREHPYGRY[443]). This segment has several sparsely spread basic residues and contains two nuclear targeting motifs: PPXXR and RXHPYQ/GR. The RXHPYQ/GR motif, originally identified and mapped in QKI-5, is also conserved in other QKI homologs as well as mammalian Sam68, SLM1, SLM2 and nonSTAR protein *Drosophila* HNF-4 homolog; all of which have a nuclear localization signal.[44] The arginines at both ends of the RXHPYQ/GR motif are essential for nuclear localization as replacing them with alanines abolishes nuclear targeting of GFP-QKI-5.

Sam68 localizes in nuclear foci called Sam68/SLM nuclear bodies (SNBs). These dynamic structures are unique, measuring > 1 μM in diameter and are adjacent to nucleoli.[45] SNBs disassemble during mitosis and upon treatment with transcription inhibitors. SNBs are distinct from other specialized subnuclear structures such as the PML nuclear bodies, interchromatin granules or speckles and Cajal bodies.[46] SNBs are observed in immortalized and transformed cells but absent in normal cells. In general, SNB prevalence correlates with the differentiation status and tumorigenicity of cancer lines such as BT-20, Hs 578T and MCF-7 cells.[45] In cells with SNBs, there is a correlation with the existence of a large Sam68 complex with a mass >1 MDa that is composed of ~40 proteins, many of which are RNA binding proteins.[47] BT-20 cells that are poorly differentiated and highly tumorigenic in nude mice display over 90 % SNB prevalence compared to nontumorigenic Hs 587T cells with about 50% SNB prevalence. However, only 5% of the well-differentiated MCF-7 cells contain SNBs. In HeLa cells, endogenous or transfected Sam68 shows a diffuse nucleoplasmic staining with several SNBs. Sam68 mutant proteins containing a deletion within the KH domain loops 1 and 4 were sufficient to concentrate Sam68 exclusively in SNBs, indicating the importance of the KH domain in SNB organization.[45] SNBs have also been shown to contain the tyrosine kinase BRK, also called Src-like Intestinal Kinase in mice (SIK),[12] the alternative splicing factor YT521-B[48] and hnRNP A1 interacting protein (HAP), a multifunctional protein involved in RNA metabolism.[49] In heat-shocked HeLa cells, HAP colocalizes with SNBs in the so called "stressed-induced SNBs" and recruits

SRp30c and 9G8 proteins, splicing factors of the SR family.[49] Although the function of SNBs is unknown, their predominant presence in highly transformed cells may serve as a marker for the host cancer cells. In addition, the colocalization within SNBs of RNA processing factors such as STAR proteins, HAP and splicing factors suggest a role of SNBs in RNA metabolism.

Although Sam68 is predominantly nuclear, there are numerous circumstances where it has been shown to be cytoplasmic. Sam68 localizes in the soma and dendrites of hippocampal neurons during depolarization,[50] where it associates with a number of plasticity-related mRNAs including the translation elongation factor eEF1A.[36] Sam68 is localized to the cytoplasm of spermatocytes during meiosis where it promotes the translation of a subset of mRNAs.[51] Total internal reflection fluorescence (TIRF) microscopy was used to observe Sam68 near the plasma membrane during cell attachment where it regulates the activity of Src by associating with c-src tyrosine kinase (Csk).[11] The presence of RNA binding proteins near the plasma membrane has been revealed by using quantitative mass spectrometry and defining a new structure termed spreading initiation centers (SICs).[52] Cells differentially labeled with light and heavy (deuterium) amino acids in suspension and cells attached on fibronectin, respectively, were lysed and cell lysates utilized to immunoprecipitate vinculin and talin complexes. The components were identified by mass spectrometry and the major category of proteins identified were RNA binding proteins including Sam68. Messenger RNAs were differentially identified at sites of membrane attachment using microarray analysis.[53] Thus specific RNAs are present near the plasma membrane during cell attachment, suggesting that RNA binding proteins are likely to also participate in RNA regulation near the plasma membrane.

In response to extracellular stress, Sam68 localizes to stress granules within the cytoplasm. Sam68 also accumulates in cytoplasmic granules with poliovirus infection, urate crystals and oxidative stress.[54-56] In addition, Sam68 mutant proteins are known to localize to cytoplasmic granules.[45,56,57] Actually Sam68 with C-terminal deletions are restricted to the cytoplasm and have been shown to function as dominant inhibitors of HIV-1 replication.[56-58] Therefore, manipulating Sam68 function may be a means to prevent HIV replication.[56,59,60]

Sam68 SIGNALING MOTIFS

The key feature of STAR proteins is the presence of various motifs that mediate protein-protein interactions and consensus sequences for enzymes that add post-translational modifications. Sam68 contains 6 polyproline rich sequence motifs that are the sites of interaction with numerous SH3 and WW domain binding proteins.[62] Sam68 has been observed to interact with the SH3 domains of Src kinases, SIK/BRK, PI3K p85α, PLCγ-1, PRMT2, Grb-2, Grap, Itk/Tec/BTK, Nck and Vav (reviewed in ref. 10 and 61). The Sam68 proline motif P3 and P4 are sites of interactions with the formin binding proteins FBP21 and FBP30 WW domains. The association with WW domain containing proteins likely represents a nuclear function of Sam68.

The STAR proteins Sam68, SLM1, SLM2 and QKI have tyrosine-rich C-termini which are sites of tyrosine phosphorylation. Sam68 is tyrosine phosphorylated by numerous soluble tyrosine kinases including $p60^{src}$, $p59^{fyn}$, $p56^{lck}$, ZAP-70 and SIK/BRK. QKI and SLM1 have been shown to be substrates of $p60^{src}$ and $p59^{fyn}$. However, the tyrosine kinases that phosphorylate SLM2 remain unknown. There are reports of several cell surface receptors that induce the tyrosine phosphorylation of Sam68. Insulin, leptin and ligation of CD16, CD32 and T-cell receptors have been observed to increase the tyrosine phosphorylation of Sam68.[63] Tyrosine phosphorylated Sam68 leads to the association of Sam68 with numerous SH2 domain containing proteins including Src family kinases, SIK/BRK, Grb2, Grap, Nck, PLCγ-1, RasGAP, PI3K p85α, Itk/Tec and Csk family kinases (reviewed in refs. 10 and 11). These observations are consistent with the fact that Sam68 functions as an adapter protein.[11,64] It is known that each SH2 domain has its own specificity. However, determining the sites of SH2 domain binding in Sam68 is difficult as it contains 16 tyrosines in 50 amino acids and the tyrosines are often clustered together. In addition, the absence of lysine and arginines in the C-terminus of Sam68 has prevented traditional tryptic mapping analyses. The best way to study the phosphorylation is by using phosphospecific antibodies which have been generated for Sam68 tyrosines 435, 440 and 443.[12]

Sam68 has also been shown to be lysine acetylated by the acetyltransferase CBP in its N-terminal region flanking the KH domain. Elevated acetylated Sam68 is observed in certain tumorigenic breast cancer cell lines and correlates with increased association of RNA as assessed by poly (U) binding.[65] Sam68 was also shown to be sumoylated by PIAS1 at lysine 96 and this influences the ability of Sam68 to regulate apoptosis and regulate transcription.[66]

ARGININE METHYLATION

Many RNA binding proteins including STAR proteins contain glycine-arginine rich (GAR) motifs and RGG boxes. The latter is a known as a RNA binding domain as exemplified by FMRP where the RGG boxes recognize RNA that contains G quartets.[67] Arginines within GAR motifs and RGG boxes are known sites of arginine methylation by protein arginine methyltransferases.[68] Protein arginine methyltransferases (PRMTs) catalyze the sequential transfer of methyl groups from S-adenosyl-L-methionine to the guanidino nitrogen atoms of certain arginine residues. The guanidino nitrogens normally favor hydrogen bonding and Van der Waal contacts which is disrupted in the presence of sterically hindering methyl group(s). The arginine methylation of GAR motifs in RNA binding proteins has been shown to alter protein-protein interactions, protein-RNA interactions and protein localization.[68] STAR proteins Sam68, SLM1, SLM2, GRP33, QKI5 and the original KH domain containing protein, hnRNPK, are methylated in vivo.[69] In Sam68, RG sequences flank proline-rich motifs P0 (RQPPLPHRGGGGSRG), P3 (RGRGAA PPPPPVPRGRG) and P4 (RGVPPPPTVRG). In vivo assays confirm that these arginines are indeed methylated by PRMT1. Moreover, mass spectrometry revealed that Sam68 Arg 45 and Arg 52 within proline motif P0 region and Arg 304 within the P3 region are dimethylated, whereas Arg 310, 315, 320 and 325 present in P4 undergo monomethylation.[69] Overall, at least half of the 14 arginine residues within the RG-rich cluster are methylated.[70] Methylation of Sam68 is markedly reduced in PRMT1[-/-] ES cells, confirming that PRMT1 is the major enzyme catalyzing the methylation of Sam68 in vivo.[69] Hypomethylated Sam68 is mislocalized[69] and displays defective protein-protein interactions.[62] Thus, Sam68 arginine methylation negatively regulates SH3, but not WW domain interactions. The prevention of interaction with cytoplasmic SH3 domain-containing proteins suggests a mechanism by which the cytoplasmic functions of Sam68 are switched to predominantly nuclear functions .

STAR PROTEIN MOUSE MODELS

Yeast, *C. elegans*, *Drosophila*, *Xenopus* and mouse models are powerful genetic models to assess the physiological role of genes. The roles of Sam68 and QKI established from mouse models are discussed.

Sam68 NULL MICE

Sam68-deficient mice were generated by the targeted disruption of *sam68* exons 4 and 5 encoding most of the KH domain.[71] Heterozygous Sam68[+/-] mice are normal. At birth, few Sam68[-/-] pups survived the first day and the reason for this is unknown. However, the few Sam68[-/-] mice that did survive the perinatal period lived a normal lifespan with no major illnesses or visible defects. However, Sam68[-/-] males are infertile and the Sam68[-/-] females have difficulty caring for their young.[35,71,72] Moreover, the Sam68[-/-] mice weigh significantly less than their heterozygous and wild-type littermates.[73] A battery of behavior tests were performed and the Sam68[-/-] mice exhibited behavior abnormalities with motor coordination defects, as assessed by beam walking and rotorod performance.[73] Similar to *Quaking* mice, these findings support a role for Sam68 in the central nervous system (CNS) in the regulation of motor coordination.

The Src tyrosine kinase is known to play a role in bone metabolism[74,75] and since Sam68 is a substrate of Src, Sam68 deficient mice were assessed for skeletal abnormalities. Mammals are known to lose bone mass with age and 12 month-old wild-type mice have a decrease of ~75% in bone mass compared with younger 4 month-old wild-type mice.[72] However, the bone mass of Sam68[-/-] mice was preserved with age. In fact, the bone volume of the 12 month-old Sam68[-/-] mice was virtually indistinguishable from that of 4 month-old wild-type or Sam68[-/-] mice suggesting that old Sam68[-/-] mice maintain a "young" bone phenotype. This was confirmed histologically. It was also noted that bone marrow stromal cells derived from Sam68[-/-] mice have a differentiation advantage for the osteogenic pathway compared to their wild-type counterparts. [72] As bone marrow cells have the potential to differentiate into the osteogenic or adipogenic pathway, it was shown that the mouse embryo fibroblasts (MEFs) derived from Sam68[-/-] mice were compromised in their ability to differentiate into adipocytes.[72] Furthermore, in vivo it was observed that sections of bones from 12 month-old Sam68[-/-] mice had fewer bone marrow adipocytes compared with their age-matched wild-type littermate controls. These findings identify Sam68 as a regulator of bone marrow mesenchymal stem cell fate.[72] A role for Sam68 in mesenchymal stem cells is consistent with the observation that Sam68 was identified as a key protein in the differentiation of neural stem cells using a gene trap strategy.[43]

A striking phenotype observed in Sam68 deficient mice is the fact that the males are infertile[71] and contain severe defects in spermatogenesis.[35] The males generate few spermatozoa and the ones that are generated exhibit dramatic motility defects and were unable to fertilize eggs. Microarray analysis revealed over 400 genes that are differentially expressed in testis between wild-type and Sam68 deficient mice.[35] Sam68 is required for polysomal recruitment of specific mRNAs during spermatogenesis, demonstrating a physiological role for Sam68 in mRNA translation[35,51] (see Ehrmann and Elliott chapter for details). A similar function in depolarized neurons has been proposed.[36,50] A role in escorting and 'marking' viral RNAs out of the cytoplasm was also suggested for Sam68.[76]

Sam68 deficient mice were bred with the mammary-targeted polyoma middle T antigen oncogene (MMTV-PyMT) transgenic mice[77] to examine whether the loss of Sam68 affected mammary tumor onset. Interestingly, Sam68 haploinsufficiency impedes mammary tumor onset in vivo driven by the potent PyMT oncogene.[72] The effect was cell-autonomous as the Sam68 knockdown in PyMT-transformed cell lines also delayed tumorigenesis and metastasis formation in *nude* mice. Interestingly, tumor extracts isolated from PyMT/Sam68[+/-] mice compared with PyMT/Sam68[+/+] mice contained activated Src kinases suggesting that Sam68 may negatively regulate Src kinase activity.[72] It was also observed that Sam68 deficient MEFs exhibited sustained Src activity after cell attachment resulting in the constitutive tyrosine phosphorylation and activation of p190RhoGAP.[11] As a result, Sam68-deficient MEFs exhibited cell migration defects, as a consequence of deregulated RhoA and Rac1 activity.[11] By total internal reflection fluorescence (TIRF) microscopy, Sam68 was localized near the plasma membrane after cell attachment coinciding with its tyrosine phosphorylation at its C-terminal tyrosines and association with Csk. These findings showed that Sam68 localizes near the plasma membrane during cell attachment and serves as an adaptor protein to modulate Src activity for proper signaling to small Rho GTPases.[11] These findings demonstrate that Sam68 is a modulator of Src tyrosine kinase activity in vivo and a signaling requirement for tumorigenesis and metastasis.

QKI MOUSE MODELS

The homozygous *quaking* viable (*Qk[v]*) mice, initially described in 1964, display vigorous tremors in their hind limbs starting around postnatal day 10 (P10) and develop tonic clonic seizures as adults (for review see ref. 78). *Qk[v]* mice have pronounced dysmyelination in the CNS and peripheral nervous systems (PNS). This dysmyelination phenotype results from reduced numbers of mature oligodendrocytes in the CNS, defects in Schwann cell maturation in the PNS, reduced number of myelin lamellae produced and the failure of the resulting myelin to properly compact.[78-80] The myelination defects were studied extensively for years, but the molecular defect was only uncovered in 1996.[81] The *QKI* gene is located on human chromosome 6q26-q27 and encodes three major isoforms, QKI5, QKI6 and QKI7.[81] These isoforms are alternatively spliced at the C-terminus and the QKI5 protein harbors a nuclear localization signal that is absent in QKI6 and QKI7.[44,82] The genetic defect of the *Qk[v]* is a 1MB deletion encompassing part of the *Qk* promoter[81] preventing the expression of QKI6 and QKI7 isoforms in myelinating cells, but not astrocytes.[83] The proper expression of QKI6 and QKI7 has been shown to contribute to the many observed defects in RNA metabolism in oligodendrocytes and Schwann cells.[78]

A *Qk* null allele has been generated by homologous recombination in mice and the embryos die at E9.5-10.5 with defects in blood vessel and neural tube formation.[84] Four *Qk* mutant alleles were generated with N-ethyl-N-nitrosourea (ENU) and these also exhibit embryonic lethality consistent with these mutants resulting in the complete loss-of-function of QKI.[85,86] There exists a viable ENU induced viable *Qk* allele (*Qk[e5]*) that displays a phenotype similar to *Qk[v]* mice but much more severe.[87] The reason for the viability of the *Qk[v]* and *Qk[e5]* mice is likely the result of the maintenance of the QKI5 isoform during embryogenesis (see Justice and Hirschi chapter for more detail). A QKI6 expressing transgenic allele under the control of the proteolipid protein (PLP) promoter expresses the isoform in a glial-specific manner and is able to rescue the dysmyelination phenotype of the Qk^v mice.[88]

In mice, the peak expression of the QKI6 and QKI7 isoforms coincides with myelination at postnatal day 14.[81] The observed defects in oligodendrocyte and Schwann cell differentiation and myelination of Qk^v mice is caused by improper QKI6 and QKI7 expression.[89,90] Indeed, the over-expression of QKI6 and QKI7 promote glial cell differentiation, at least in part, through the stabilization of the p27^{KIP1} mRNA.[89] It is known that cellular differentiation initially requires cell cycle arrest followed by changes in gene expression. p27^{KIP1} is a cyclin-dependent kinase (CDK) inhibitor that binds to and prevents the activation of cyclin E-CDK2 or cyclin D-CDK4 complexes and thus controls the cell cycle progression at G_1 phase.[91] $p27^{KIP1}$ accumulation is required for oligodendrocyte precursor differentiation.[92-95] When QKI6/7 are expressed, the $p27^{KIP1}$ mRNA and subsequently the protein accumulates causing cell cycle arrest and then differentiation.[89,90] In addition to regulating glial cell fate, the ectopic expression of QKI6 and QKI7 are able to direct the progenitor cells of the ventricular zone to migrate to areas of high myelination, such as the corpus callosum and become oligodendrocytes.[89] In vivo, progenitors expressing QKI6 and QKI7 also developed into astrocytes, migrated and localized to the border of the corpus callosum into the rostral marginal zone.[89]

The QKI proteins are sequence-specific RNA binding proteins. The QKI recognition element (QRE) was defined as direct repeats of the hexanucleotide 5'-ACUAAY-3' and a half-site 5'-UAAY-3' spaced by 1 to 20 nucleotides.[29,32] A bioinformatics analysis identified 1433 putative mRNA targets involved in development, cell adhesion and cell differentiation.[29] The mRNAs that contain a QRE include $p27^{KIP1}$, $Krox20$ and myelin basic proteins (MBP). The MBP mRNA stability and nuclear retention were observed in Qk^v mice resulting in severe hypomyelination.[96,97]

STAR PROTEINS AND HUMAN DISEASES

Since STAR proteins are key components in RNA metabolism, regulating the temporal, spatial and functional dynamics of RNAs; altering their expression should have major implications for human disorders. In this section, the current literature that links QKI, Sam68 and SLM2 to complex human disorders is discussed.

OSTEOPOROSIS

A physiological role for RNA metabolism in human bone metabolism is implied from the Sam68-deficient mice.[71] Aberrant alternative splicing of collagen, Type I, alpha 1 (COLA1) and CD44 antigen has been linked with osteoporosis.[98,99] Thus, strategies that can inhibit Sam68 RNA binding should prevent age-related bone loss. Furthermore, altered Sam68 expression, polymorphisms, or mutations within the human $sam68$ gene may influence development of bone marrow adipocyte accumulation and osteoporosis.

SCHIZOPHRENIA

It was observed that myelin and oligodendrocyte defects may contribute to the development of schizophrenia.[100] A decreased oligodendrocyte density in the white matter of schizophrenia patients and other alterations in schizophrenia brains resemble those observed in the Qk^v mice.[101] A schizophrenia susceptibility locus was mapped to chromosome 6q25-6q26, the location of the QKI gene.[101-103] However, seven genetic markers around the QKI promoter region were not associated with schizophrenia in 288 individuals diagnosed with schizophrenia in a Chinese study.[104] Several studies have examined the expression of various transcripts in areas of the brain affected by schizophrenia such as the anterior cingulated cortex and the superior temporal cortex. Several myelin-related genes were decreased in the white matter of schizophrenia patients including the mRNA encoding myelin-associated glycoprotein (MAG),[102,105] proteolipid protein 1 ($PLP1$) and QKI.[101-103] Many of these transcripts are identified targets of the QKI proteins and the differences observed can be explained by variation in the relative mRNA levels of QKI. These findings suggest more studies are required to firmly link QKI to the pathology of schizophrenia.

ATAXIA

There currently exists no genetic association between STAR domain proteins and ataxia in humans. Studies using Qk^v and Sam68 deficient mice do provide evidence, however, linking STAR proteins to ataxia and cerebellar defects. The QKI isoforms link ataxia with the Qk^v and Qk^{e5} mice and show Purkinje cell axonal swelling, indicative of neuronal degeneration.[87,106] Moreover, a protein-protein interaction map was generated for genes involved in ataxia and, interestingly, many RNA binding proteins including QKI were part of this interactome.[107]

Sam68 is highly expressed in the brain especially the cerebellum. Sam68-null mice exhibit motor coordination defects, as assessed by beam walking and rotorod performance.[73] The Sam68-null mice exhibit more hind paw faults in beam walking tests and fell from the rotating drum at lower speeds when compared to their wild-type controls. These findings support a role for Sam68 in the CNS in the regulation of motor coordination.[73] Actually several splicing and RNA gain-of-function defects have been linked to ataxia in humans.[23,108]

CANCER

RNA binding proteins are known regulators of key events in RNA metabolism influencing the cell cycle, proliferation and cellular migration.[23] Therefore, it is not surprising that RNA binding proteins have intimate links with cellular transformation, cancer initiation and progression. The fact that Sam68 is a known substrate of Src family kinases and BRK,[109] suggests that the tyrosine phosphorylation of Sam68 may be required for the maintenance of tumors. Interestingly, the BRK tyrosine kinase is not expressed in normal mammary gland, but is overexpressed in over 60% of primary breast tumors.[110,111] Actually using 426 archival breast cancer samples, the BRK expression was of significant prognostic value and correlated with the expression of PTEN, MAPK, p-MAPK and Sam68.[112,113] Consistent with these findings, Sam68 is tyrosine phosphorylated in many human breast tumors and cell lines[12,65] and this regulates its complex composition and function.[47] The elevated expression of Sam68 has been observed in 35% of prostate cancers examined in a small cohort and its expression was necessary to maintain cell survival of the LNCaP prostate cell line.[114] *Sam68* mRNA and protein expression was elevated in 241 renal cell carcinoma cell lines and tissues examined. The elevated expression and the cytoplasmic localization of Sam68 correlated with poor prognosis in renal cell carcinomas.[115] These data suggest that strategies that neutralize Sam68 function in vivo should be of therapeutic value for cancer treatment. This statement is reinforced by data in Sam68 deficient mice showing that PyMT-driven mammary tumorigenesis is delayed compared to wild type mice.[72]

Gliomas represent > 70% of all brain tumors and the overall prognosis of treatment success is poor with a mean survival of 12.2 to 18.2 months after diagnosis and a 5-year survival rate that ranges 42 to 92%.[116] Glioblastoma multiforme (GBM) often harbor alterations affecting the p53 and retinoblastoma pathways, amplification of the epidermal growth factor receptor, deletion of PTEN and deletion of chromosome 6q26-27.[116] The *QKI* gene spans over 159 kb on human chromosome 6p26 and, not surprisingly, it has been identified to be deleted in several GBMs.[117-119] In addition, alterations in QKI isoform expression is observed in ~30% (6/20) of human GBM, whereas QKI isoform expression was unchanged in all of the Schwannomas and Meningiomas tested.[120] As the majority of GBMs also show alterations in genes such as p53 and retinoblastoma,[118] amplifications and/or rearrangements of EGFR, or loss of wild-type PTEN; it remains to be determined if *QKI* alteration leads to the initiation or the progression of GBM.

CONCLUSION

The Sam68 deficient mice and *Qk^v* mice have provided important information about the physiological role of STAR proteins. The challenge ahead will be to define the regulation of STAR proteins, their RNA targets and the molecular function(s) of individual isoforms. In addition, the knowledge of the composition of STAR RNP complexes and the development of animal models that closely resembles human diseases will facilitate the development of novel therapeutics for the associated particular disease.

ACKNOWLEDGEMENTS

The work on Sam68 was supported by a grant from the Canadian Institutes of Health Research (CIHR) MT-13377 and work on QKI was supported by a grant from the Multiple Sclerosis Society of Canada.

REFERENCES

1. Di Fruscio M, Chen T, Richard S. Two novel Sam68-like mammalian proteins SLM-1 and SLM-2: SLM-1 is a Src substrate during mitosis. Proc Natl Acad Sci USA 1999; 96:2710-2715.
2. Venables JP, Vernet C, Chew SL et al. T-STAR/ETOILE: a novel relative of Sam68 that interacts with an RNA-binding protein implicated in spermatogenesis. Hum Mol Genetics 1999; 8:959-969.
3. Wong G, Muller O, Clark R et al. Molecular cloning and nucleic acid binding properties of the GAP-associated tyrosine phosphoprotein p62. Cell 1992; 69(3):551-558.

4. Vogel LB, Fujita DJ. p70 phosphorylation and binding to p56lck is an early event in interleukin-2 induced onset of cell cycle progression in T-lymphocytes. J Biol Chem 1995; 270:2506-2511.
5. Fumagalli S, Totty NF, Hsuan JJ et al. A target for Src in mitosis. Nature 1994; 368:871-874.
6. Taylor SJ, Shalloway D. An RNA-binding protein associated with src through its SH2 and SH3 domains in mitosis. Nature 1994; 368:867-871.
7. Richard S, Yu D, Blumer KJ et al. Association of p62, a multi-functional SH2- and SH3-binding protein, with src-family tyrosine kinases, Grb2 and phospholipase Cγ-1. Mol Cell Biol 1995; 15:186-197.
8. Weng A, Thomas SM, Rickles RJ et al. Identification of Src, Fyn and Lyn SH3-binding proteins: Implications for a function of SH3 domains. Mol Cell Biol 1994; 14:4509-4521.
9. Lock P, Fumagalli S, Polakis P et al. The human p62 cDNA encodes Sam68 and not the RasGAP-associated p62 protein. Cell 1996; 84(1):23-24.
10. Lukong KE, Richard S. Sam68, the KH domain-containing superSTAR. Biochim Biophys Acta 2003; 1653:73-86.
11. Huot ME, Brown CM, Lamarche-Vane N et al. An adaptor role for cytoplasmic Sam68 in modulating Src activity during cell polarization. Mol Cell Biol 2009; 29:1933-1943.
12. Lukong KE, Larocque D, Tyner AL et al. Tyrosine phosphorylation of sam68 by breast tumor kinase regulates intranuclear localization and cell cycle progression. J Biol Chem 2005; 280:38639-38647.
13. Paronetto MP, Achsel T, Massiello A et al. The RNA-binding protein Sam68 modulates the alternative splicing of Bcl-x. J Cell Biol 2007; 176:929-939.
14. Sanchez-Margalet V, Gonzalez-Yanes C, Najib S et al. The expression of Sam68, a protein involved in insulin signal transduction, is enhanced by insulin stimulation. Cell Mol Life Sci 2003; 60(4):751-758.
15. Lang V, Mege D, Semichon M et al. A dual participation of ZAP-70 and scr protein tyrosine kinases is required for TCR-induced tyrosine phosphorylation of Sam68 in Jurkat T-cells. Eur J Immunol 1997; 27(12):3360-3367.
16. Jabado N, Jauliac S, Pallier A et al. Sam68 association with p120GAP in CD4+ T-cells is dependent on CD4 molecule expression. J Immunol 1998; 161:2798-2803.
17. Martin-Romero C, Sanchez-Margalet V. Human leptin activates PI3K and MAPK pathways in human peripheral blood mononuclear cells: possible role of Sam68. Cell Immunol 2001; 212:83-91.
18. Medema JP, Pronk GJ, de Vries-Smits AM et al. Insulin-induced tyrosine phosphorylation of a M(r) 70,000 protein revealed by association with the Src homology 2 (SH2) and SH3 domains of p120GAP and Grb2. Cell Growth Differ 1996; 7(4):543-550.
19. Sanchez-Margalet V, Najib S. Sam68 is a docking protein linking GAP and PI3K in insulin receptor signaling. Mol Cell Endocrinol 2001; 183(1-2):113-121.
20. Di Fruscio M, Chen T, Bonyadi S et al. The identification of two Drosophila K homology domain proteins. Kep1 and SAM are members of the Sam68 family of GSG domain proteins. J Biol Chem. Nov 13 1998; 273(46):30122-30130.
21. Chen T, Damaj BB, Herrera C et al. Self-association of the single-KH-domain family members Sam68, GRP33, GLD-1 and Qk1: role of the KH domain. Mol Cell Biol 1997; 17(10):5707-5718.
22. Glisovic T, Bachorik JL, Yong J et al. RNA-binding proteins and post-transcriptional gene regulation. FEBS Lett 2008; 582:1977-1986.
23. Lukong KE, Chang KW, Khandjian EW et al. RNA-binding proteins in human genetic disease. Trends Genet 2008; 24:416-425.
24. Valverde R, Edwards L, Regan L. Structure and function of KH domains. FEBS J 2008; 275:2712-2726.
25. García-Mayoral MF, Díaz-Moreno I, Hollingworth D et al. The sequence selectivity of KSRP explains its flexibility in the recognition of the RNA targets. Nucleic Acids Res 2008; 36:5290-5296.
26. Chen T, Richard S. Structure-function analysis of Qk1: a lethal point mutation in mouse quaking prevents homodimerization. Mol Cell Biol 1998; 18(8):4863-4871.
27. Lin Q, Taylor SJ, Shalloway D. Specificity and determinants of Sam68 RNA binding. Implications for the biological function of K homology domains. J Biol Chem 1997; 272(43):27274-27280.
28. Galarneau A, Richard S. The STAR RNA binding proteins GLD-1, QKI, SAM68 and SLM-2 bind bipartite RNA motifs. BMC Mol Biol 2009; 10:47.
29. Galarneau A, Richard S. Target RNA motif and target mRNAs of the Quaking STAR protein. Nat Struct Mol Biol 2005; 12:691-698.
30. Arning S, Gruter P, Bilbe G et al. Mammalian splicing factor SF1 is encoded by variant cDNAs and binds to RNA. RNA 1996; 2:794-810.
31. Berglund JA, Chua K, Abovich N et al. The splicing factor BBP interacts specifically with the pre-mRNA branch-point sequence UACUAAC. Cell 1997; 89:781-787.
32. Ryder SP, Frater L, A et al. RNA target specificity of the STAR/GSG domain post-transcriptional regulatory protein GLD-1. Nat Struct Mol Biol 2004; 11:20-28.
33. Itoh M, Haga I, Li Q-H et al. Identification of cellular mRNA targets for RNA-binding protein Sam68. Nucl. Acids Res 2002; 30:5452-5464.

34. Tremblay GA, Richard S. mRNAs associated with the Sam68 RNA binding protein. RNA Biology 2006; 3:1-4.
35. Paronetto MP, Messina V, Bianchi E et al. Sam68 regulates translation of target mRNAs in male germ cells, necessary for mouse spermatogenesis. J Cell Biol 2009; 185:235-249.
36. Grange J, Belly A, Dupas S et al. Specific interaction between Sam68 and neuronal mRNAs: Implication for the activity-dependent biosynthesis of elongation factor eEF1A. J Neurosci Res 2009; 87:12-25.
37. Matter N, Herrlich P, Konig H. Signal-dependent regulation of splicing via phosphorylation of Sam68. Nature 2002; 420:691-695.
38. Batsche E, Yaniv M, Muchardt C. The human SWI/SNF subunit Brm is a regulator of alternative splicing. Nat Struct Mol Biol 2006; 13:22-29.
39. Cheng C, Sharp PA. Regulation of CD44 alternative splicing by SRm160 and its potential role in tumor cell invasion. Mol Cell Biol 2006; 26(1):362-370.
40. Rajan P, Gaughan L, Dalgliesh C et al. The RNA-binding and adaptor protein Sam68 modulates signal-dependent splicing and transcriptional activity of the androgen receptor. J Pathol 2008; 215:67-77.
41. Maroni P, Citterio L, Piccoletti R et al. Sam68 and ERKs regulate leptin-induced expression of OB-Rb mRNA in C2C12 myotubes. Mol Cell Endocrinol 2009; 309:26-31.
42. Chawla G, Lin CH, Han A et al. Sam68 regulates a set of alternatively spliced exons during neurogenesis. Mol Cell Biol 2008; 29:201-213.
43. Moritz S, Lehmann S, Faissner A et al. An induction gene trap screen in neural stem cells reveals an instructive function of the niche and identifies the splicing regulator sam68 as a tenascin-C-regulated target gene. Stem Cells 2008; 26:2321-2331.
44. Wu J, Zhou L, Tonissen K et al. The quaking I-5 (QKI-5) has a novel nuclear localization signal and shuttles between the nucleus and the cytoplasm. J Biol Chem 1999; 274:29202-29210.
45. Chen T, Boisvert FM, Bazett-Jones DP et al. A role for the GSG domain in localizing Sam68 to novel nuclear structures in cancer cell lines. Mol Biol Cell 1999; 10(9):3015-3033.
46. Spector DL. Nuclear domains. J Cell Sci 2001; 114:2891-2893.
47. Huot ME, Vogel G, Richard S. Identification of a Sam68 ribonucleoprotein complex regulated by epidermal growth factor. J Biol Chem 2009; [Epub ahead of print].
48. Hartmann AM, Nayler O, Schwaiger FW et al. The interaction and colocalization of Sam68 with the splicing-associated factor YT521-B in nuclear dots is regulated by the Src family kinase p59fyn. Mol Biol Cell 1999; 10:3909-3926.
49. Denegri M, Chiodi I, Corioni M et al. Stress-induced nuclear bodies are sites of accumulation of pre-mRNA processing factors. Mol Biol Cell 2001; 12:3502-3514.
50. Grange J, Boyer V, Fabian-Fine R et al. Somatodendritic localization and mRNA association of the splicing regulatory protein Sam68 in the hippocampus and cortex. J Neurosci Res 2004; 75:654-666.
51. Paronetto MP, Zalfa F, Botti F et al. The nuclear RNA-binding protein Sam68 translocates to the cytoplasm and associates with the polysomes in mouse spermatocytes. Mol Biol Cell 2006; 17:14-24.
52. de Hoog CL, Foster LJ, Mann M. RNA and RNA binding proteins participate in early stages of cell spreading through spreading initiation centers. Cell 2004; 117:649-662.
53. Mili S, Moissoglu K, Macara IG. Genome-wide screen reveals APC-associated RNAs enriched in cell protrusions. Nature 2008; 453:115-119.
54. McBride AE, Schlegel A, Kirkegaard K. Human protein Sam68 relocalization and interaction with poliovirus RNA polymerase in infected cells. Proc Natl Acad Sci USA 1996; 93:2296-2301.
55. Gilbert C, Barabe F, Rollet-Labelle E et al. Evidence for a role for SAM68 in the responses of human neutrophils to ligation of CD32 and to monosodium urate crystals. J Immunol 2001; 166:4664-4671.
56. Henao-Mejia J, Liu Y, Park IW et al. Suppression of HIV-1 Nef translation by Sam68 mutant-induced stress granules and nef mRNA sequestration. Mol Cell 2009; 33:87-96.
57. Soros V, Valderamma H, Richard S et al. Inhibition of HIV-1 Rev function by a dominant negative mutant of Sam68 through sequestration of unspliced RNA at perinuclear bundles. J Virology 2001; 75:8203-8215.
58. Reddy TR, Xu W, Mau JK et al. Inhibition of HIV replication by dominant negative mutants of Sam68, a functional homolog of HIV-1 Rev. Nat Med 1999; 5(6):635-642.
59. Suhasini M, Reddy TR. Cellular proteins and HIV-1 Rev function. Curr HIV Res 2009; 7:91-100.
60. Cochrane A. Inhibition of HIV-1 gene expression by Sam68 Delta C: multiple targets but a common mechanism? Retrovirology 2009; 6:22.
61. Lazer G, Pe'er L, Schapira V et al. The association of Sam68 with Vav1 contributes to tumorigenesis. Cell Signal 2007; 19:2479-2486.
62. Bedford MT, Frankel A, Yaffe MB et al. Arginine methylation inhibits the binding of proline-rich ligands to Src homology 3, but not WW, domains. J Biol Chem 2000; 275:16030-16036.
63. Najib S, Martin-Romero C, Gonzalez-Yanes C et al. Role of Sam68 as an adaptor protein in signal transduction. Cell Mol Life Sci 2005; 62(1):36-43.

64. Cheung N, Chan LC, Thompson A et al. Protein arginine-methyltransferase-dependent oncogenesis. Nat Cell Biol 2007; 9:1208-1215.
65. Babic I, Jakymiw A, Fujita DJ. The RNA binding protein Sam68 is acetylated in tumor cell lines and its acetylation correlates with enhanced RNA binding activity. Oncogene 2004; 23:3781-3789.
66. Babic I, Cherry E, Fujita DJ. SUMO modification of Sam68 enhances its ability to repress cyclin D1 expression and inhibits its ability to induce apoptosis. Oncogene 2006; 25:4955-4964.
67. Darnell JC, Jensen KB, Jin P et al. Fragile X mental retardation protein targets G quartet mRNAs important for neuronal function. Cell 2001; 107:489-499.
68. Bedford MT, Richard S. Arginine methylation an emerging regulator of protein function. Mol Cell 2005; 18:263-272.
69. Côté J, Boisvert FM, Boulanger MC et al. Sam68 RNA binding protein is an in vivo substrate for protein arginine N-methyltransferase 1. Mol Biol Cell 2003; 14(1):274-287.
70. Rappsilber J, Friesen WJ, Paushkin S et al. Detection of Arginine Dimethylated Peptides by Parallel Precursor Ion Scanning Mass Spectrometry in Positive Ion Mode. Anal Chem 2003; 75(13):3107-3114.
71. Richard S, Torabi N, Franco GV et al. Ablation of the Sam68 RNA binding protein protects mice from age-related bone loss. 2005; 1:e74.
72. Richard S, Vogel G, Huot ME et al. Sam68 haploinsufficiency delays onset of mammary tumorigenesis and metastasis. Oncogene 2008; 27:548-556.
73. Lukong KE, Richard S. Motor coordination defects in mice deficient for the Sam68 RNA-binding protein. Behav Brain Res 2008; 189:357-363.
74. Soriano P, Montgomery C, Geske R et al. Targeted Disruption of the c-src Proto-Oncogene Leads to Osteopetrosis in Mice. Cell 1991; 64:693-702.
75. Miyazaki T, Tanaka S, Sanjay A et al. The role of c-Src kinase in the regulation of osteoclast function. Mod Rheumatol 2006; 16:68-74.
76. Coyle JH, Guzik BW, Bor YC et al. Sam68 enhances the cytoplasmic utilization of intron-containing RNA and is functionally regulated by the nuclear kinase Sik/BRK. Mol. Cell Biol 2003; 23:92-103.
77. Andrechek ER, Muller WJ. Tyrosine kinase signalling in breast cancer: tyrosine kinase-mediated signal transduction in transgenic mouse models of human breast cancer. Breast Cancer Res 2000; 2:211-216.
78. Chenard CA, Richard S. New implications for the QUAKING RNA binding protein in human disease. J Neurosci Res 2008; 86:233-242.
79. Hogan EL, Greenfield S. Animal models of genetic disorders of myelin. Myelin 1984:489-534.
80. Hardy RJ. Molecular defects of the dysmyelinating mutant quaking. J Neurosci Res 1998; 51:417-422.
81. Ebersole TA, Chen Q, Justice MJ et al. The quaking gene unites signal transduction and RNA binding in the developing nervous system. Nature Genetics 1996; 12:260-265.
82. Pilotte J, Larocque D, Richard S. Nuclear translocation controlled by alternatively spliced isoforms inactivates the QUAKING apoptotic inducer. Genes & Dev 2001; 15:845-858.
83. Hardy RJ, Loushin CL, Friedrich Jr. VL et al. Neural cell type-specific expression of QKI proteins is altered in the quaking viable mutant mice. J Neuroscience 1996; 16:7941-7949.
84. Li Z, Takakura N, Oike Y et al. Defective smooth muscle development in qkI-deficient mice. Dev Growth Differ 2003; 45(5-6):449-462.
85. Justice MJ, Bode VC. Three ENU-induced alleles of the murine quaking locus are recessive embryonic lethal mutations. Genet Res 1988; 51:95-102.
86. Cox RD, Hugill A, Shedlovsky A et al. Contrasting effects of ENU induced embryonic lethal mutations of the quaking gene. Genomics 1999; 57:333-341.
87. Noveroske JK, Hardy R, Dapper JD et al. A new ENU-induced allele of mouse quaking causes severe CNS dysmyelination. Mamm Genome 2005; 16:672-682.
88. Zhao L, Tian D, Xia M et al. Rescuing qkV dysmyelination by a single isoform of the selective RNA-binding protein QKI. J Neurosci 2006; 26:11278-11286.
89. Larocque D, Galarneau A, Liu HN et al. Protection of the p27KIP1 mRNA by quaking RNA binding proteins promotes oligodendrocyte differentiation. Nat Neurosci 2005; 8:27-33.
90. Larocque D, Fragoso G, Huang J et al. The QKI-6 and QKI-7 RNA binding proteins block proliferation and promote Schwann cell myelination. PLoS ONE 2009; 4:e5867.
91. Besson A, Dowdy SF, Roberts JM. CDK inhibitors: cell cycle regulators and beyond. Dev Cell 2008; 14:159-169.
92. Friessen AJ, Miskimins WK, Miskimins R. Cyclin-dependent kinase inhibitor p27kip1 is expressed at high levels in cells that express a myelinating phenotype. J Neurosci Res 1997; 50(3):373-382.
93. Dyer MA, Cepko CL. p27Kip1 and p57Kip2 regulate proliferation in distinct retinal progenitor cell populations. J Neurosci 2001; 21(12):4259-4271.
94. Durand B, Gao FB, Raff M. Accumulation of the cyclin-dependent kinase inhibitor p27/Kip1 and the timing of oligodendrocyte differentiation. EMBO J 1997; 16(2):306-317.

95. Casaccia-Bonnefil P, Hardy RJ, Teng KK et al. Loss of p27Kip1 function results in increased proliferative capacity of oligodendrocyte progenitors but unaltered timing of differentiation. Development 1999; 126(18):4027-4037.
96. Li Z, Zhang Y, Li D et al. Destabilization and mislocalization of the myelin basic protein mRNAs in quaking dysmyelination lacking the Qk1 RNA-binding proteins. J Neurosci 2000; 20:4944-4953.
97. Larocque D, Pilotte J, Chen T et al. Nuclear retention of MBP mRNAs in the Quaking viable mice. Neuron 2002; 36:815-829.
98. Vidal C, Cachia A, Xuereb-Anastasi A. Effects of a synonymous variant in exon 9 of the CD44 gene on pre-mRNA splicing in a family with osteoporosis. Bone 2009; 45:736-742.
99. Xia XY, Cui YX, Huang YF et al. A novel RNA-splicing mutation in COL1A1 gene causing osteogenesis imperfecta type I in a Chinese family. Clin Chim Acta 2008; 398:148-151.
100. Stewart DG, Davis KL. Possible contributions of myelin and oligodendrocyte dysfunction to schizophrenia. Int Rev Neurobiol 2004; 59:381-424.
101. Haroutunian V, Katsel P, Dracheva S et al. The human homolog of the QKI gene affected in the severe dysmyelination "quaking" mouse phenotype: downregulated in multiple brain regions in schizophrenia. Am J Psychiatry 2006; 163:1834-1837.
102. Aberg K, Saetre P, Jareborg N et al. Human QKI, a potential regulator of mRNA expression of human oligodendrocyte-related genes involved in schizophrenia. Proc Natl Acad Sci USA 2006; 103:7482-7487.
103. Aberg K, Saetre P, Lindholm E et al. Human QKI, a new candidate gene for schizophrenia involved in myelination. Am J Med Genet B Neuropsychiatr Genet 2006; 141B:84-90.
104. Huang K, Tang W, Xu Z et al. No association found between the promoter variations of QKI and schizophrenia in the Chinese population. Prog Neuropsychopharmacol Biol Psychiatry 2009; 33:33-36.
105. McCullumsmith RE, Gupta D, Beneyto M et al. Expression of transcripts for myelination-related genes in the anterior cingulate cortex in schizophrenia. Schizophr Res 2007; 90:15-27.
106. Suzuki K, Zagoren JC. Variations of Schmidt-Lanterman incisures in Quaking mouse. Brain Res 1976; 106(1):146-151.
107. Lim DA, Suarez-Farinas M, Naef F et al. In vivo transcriptional profile analysis reveals RNA splicing and chromatin remodeling as prominent processes for adult neurogenesis. Mol Cell Neurosci 2006; 31(1):131-148.
108. Cooper TA, Wan L, Dreyfuss G. RNA and disease. Cell 2009; 136:777-793.
109. Derry JJ, Richard S, Carvajal HV et al. Sik (BRK) phosphorylates Sam68 in the nucleus and negatively regulates its RNA binding activity. Mol Cell Biol 2000; 20:6114-6126.
110. Barker KT, Jackson LE, Crompton MR. BRK tyrosine kinase expression in a high proportion of human breast carcinomas. Oncogene 1997; 15:799-805.
111. Serfas MS, Tyner AL. Brk, Srm, Frk and Src42A form a distinct family of intracellular Src-like tyrosine kinases. Oncol Res 2003; 13:409-419.
112. Aubele M, Auer G, Walch AK et al. PTK (protein tyrosine kinase)-6 and HER2 and 4, but not HER1 and 3 predict long-term survival in breast carcinomas. Br J Cancer 2007; 96:801-807.
113. Aubele M, Walch AK, Ludyga N et al. Prognostic value of protein tyrosine kinase 6 (PTK6) for long-term survival of breast cancer patients. Br J Cancer 2008; 99:1089-1095.
114. Busa R, Paronetto MP, Farini D et al. The RNA-binding protein Sam68 contributes to proliferation and survival of human prostate cancer cells. Oncogene 2007; 26(30):4372-4382.
115. Zhang Z, Li J, Zheng H et al. Expression and Cytoplasmic Localization of SAM68 Is a Significant and Independent Prognostic Marker for Renal Cell Carcinoma. Cancer Epidemiol Biomarkers Prev 2009; [Epub ahead of print].
116. Sanai N, Berger MS. Glioma extent of resection and its impact on patient outcome. Neurosurgery 2008; 62:753-764.
117. Mulholland PJ, Fiegler H, Mazzanti C et al. Genomic profiling identifies discrete deletions associated with translocations in glioblastoma multiforme. Cell Cycle 2006; 5:783-791.
118. Ichimura K, Mungall AJ, Fiegler H et al. Small regions of overlapping deletions on 6q26 in human astrocytic tumours identified using chromosome 6 tile path array-CGH. Oncogene 2006; 25:1261-1271.
119. Yin D, Ogawa S, Kawamata N et al. High-resolution genomic copy number profiling of glioblastoma multiforme by single nucleotide polymorphism DNA microarray. Mol Cancer Res 2009; 7:665-677.
120. Li ZZ, Kondo T, Murata T et al. Expression of Hqk encoding a KH RNA binding protein is altered in human glioma. Jpn J Cancer Res 2002; 93(2):167-177.

INDEX

A

Alternative splicing 4, 6, 25, 27, 28, 30, 37, 41, 55, 56, 58, 62, 69, 77, 83, 85, 99, 102, 119, 124, 133, 135, 146, 151

Apoptosis 60, 61, 69, 76, 78, 86, 93, 100, 102, 103, 107, 109, 111, 115, 148

Arginine methylation 58, 60, 73, 74, 142, 148

ASD-2 1, 2, 94, 95, 106, 109, 119, 135

Ataxia 5, 6, 142, 152

B

Blood vessel 82, 84, 85, 89, 150

Brain development 26, 27, 31, 33

Branch point binding protein (BBP) 2, 43, 123-136, 145

Breast cancer 58, 61, 148, 152

C

Caenorhabditis elegans 1, 2, 6, 12, 13, 38-40, 85, 94, 106, 107, 109, 114-116, 119, 120, 123, 135, 143, 149

Cancer 8, 26, 27, 32, 57-59, 61, 64, 120, 142, 146-148, 152, 153

Cardiovascular development 82, 88, 89

CD44 56, 59, 77, 78, 146, 151

Cell cycle 27-29, 32, 59, 61, 100, 101, 151, 152

Cell differentiation 5, 6, 27, 28, 56, 85, 95, 97, 99, 115, 146, 151

Cell growth 27, 28, 82, 86

Cell signaling 25-27, 30, 33

CLIP 78, 79

Commitment complex 124, 127, 129, 131

Consensus sequence 21, 39, 41, 43, 54, 117, 145, 147

Conservation 3, 8, 9, 39, 45, 48, 50, 74, 99

D

Drosophila 1-4, 6, 13-15, 17, 25, 38, 42, 88, 89, 93-95, 98-100, 102-104, 109, 116, 135, 143, 146, 149

Dynamics 21, 32, 47, 57, 59, 127, 146, 151

E

Embryo 4, 5, 26, 62, 74, 82-86, 88, 89, 94, 98-101, 103, 109, 111, 112, 115, 119, 149, 150

Embryogenesis 84, 93, 95, 107, 110, 113, 150

Exonic splicing enhancer (ESE) 77

Exonic splicing silencer (ESS) 77

F

Fyn   3, 4, 28-33, 57, 58, 62, 74

G

Gel mobility shift   39, 40, 42
Germ cell   56, 59, 67-71, 76, 78,
   106-110, 113-115, 117-119
Germline defective (GLD-1)   1, 2,
   6, 38-43, 48-50, 54, 55, 85, 94,
   106-111, 113-120, 135, 136, 145
Germline development   39, 106, 107,
   109, 110, 114, 115, 118-120
Germline tumor   109, 110, 114, 117
Glia   4, 5, 6, 10, 26, 27, 30, 32, 62, 89,
   94, 98, 99, 102, 103, 151
Glioma   32, 153
GYF protein   133

H

Heart   5, 82, 83, 88, 89, 98, 99
Held out wing (HOW)   1, 4, 25, 38-43,
   48-50, 63, 89, 93-104, 109, 110, 117,
   135, 136
hnRNP G   78

I

Intron   6, 7, 33, 43, 63, 77, 79, 119, 123,
   124, 126-128, 133, 134
Intronic splicing enhancer (ISE)   77
Intronic splicing silencer (ISS)   64, 77

K

Kep1   2, 93, 100, 102, 103
KH domain   3, 4, 8, 11-13, 20, 27, 39,
   43-45, 47, 50, 55, 58, 61, 70, 73, 76,
   78, 83, 84, 95, 123, 124, 142-146,
   148, 149
KHDRBS1/2/3   1, 2, 11, 42, 54, 70-72,
   79, 93, 142, 143
Kinase   3, 4, 6, 8, 10, 12, 14, 17, 25, 28,
   30, 31, 33, 57-59, 63, 74, 76, 85, 95,
   98, 100, 103, 112, 114, 126, 143,
   146-148, 149-152

M

Maternal mRNA   107, 110, 116
Meiosis   67, 69, 70, 74, 78, 117, 118,
   147
Metastasis   16, 150
Methylation   54, 55, 58, 60-62, 73, 74,
   142, 148
MicroRNA (miRNA)   1, 6-8, 10-21, 79,
   116, 117, 120
Mouse model   79, 149, 150
mRNA target   25-29, 33, 37, 43, 61, 98,
   106, 108, 110, 111, 113-117, 119,
   120, 151
mRNA translation   113
MSL5   2, 124, 126, 130-132, 135
Mud2   123-132, 134-136
Muscle   1, 5, 27, 84, 89, 94-99, 107, 115,
   119
Myelin   4, 6, 7, 14, 15, 26-33, 40, 41, 62,
   83, 85, 98, 99, 150-152
Myelination   4-6, 11, 27-32, 62, 82, 83,
   85, 150, 151
Myelin formation   26-28, 30, 31

N

NMR Spectroscopy   43
Nonsense mediated decay (NMD)   132,
   133
Nucleus   3, 4, 28, 29, 56, 58, 59, 62,
   69, 70, 76, 99, 107, 124, 127, 128,
   131-135

O

Oligodendrocyte   6, 15, 26, 29, 31, 41,
   62, 83, 84, 97-99, 150-152
Osteoporosis   142, 151
Outflow tract   82, 88, 89

P

P-body   133-135
Phosphorylation   3, 25-28, 30, 31, 33,
   54-60, 62, 63, 74, 77, 95, 103, 125,
   126, 142, 143, 147, 148, 150, 152
Post-transcriptional regulation   8, 38, 42,
   56, 62, 69

Pre-mRNA Splicing 5, 6, 67, 77, 78, 128, 130, 131, 133, 135
Prespliceosome 127, 129, 130
Protamine 76
Protein interaction 54, 55, 70, 75, 77, 78, 130, 136, 147, 148, 152
Protein-RNA interaction 38, 148
Purkinje cell 5-7, 152

Q

Quaking (QKI) 1-8, 10-12, 14, 20, 25-33, 38-44, 47-50, 55, 62, 63, 70, 82-89, 93-95, 97-99, 102, 109, 110, 119, 135, 136, 142, 143, 145-147, 149, 150-153

R

RBMX 78
Retinoic acid (RA) 59, 82, 85-87
RNA 1-6, 11, 12, 17, 20, 21, 25-30, 33, 37-50, 54-63, 69-71, 74, 77-79, 83, 85, 93, 94, 96, 99, 103, 106, 109, 110, 113, 114, 116-118, 120, 123-128, 130-136, 142, 143, 145-153
  binding protein 4, 29, 37-39, 44, 54, 55, 69-71, 77, 78, 83, 93, 103, 109, 110, 113, 117, 118, 120, 124, 127, 134, 136, 142, 143, 146-148, 151, 152
  export 124, 132, 133, 135, 136
  metabolism 1, 3-5, 11, 21, 55, 56, 63, 113, 114, 142, 146, 147, 150, 151, 152
  processing 5, 12, 17, 55, 69, 70, 79, 123, 131, 133, 134, 136, 147
  stability 135
Round spermatid 69, 70, 76

S

Saccharomyces cerevisiae 10, 17, 123
SAFB1/2 78
Sam68 1-6, 11, 20, 25, 27, 28, 33, 38-42, 48-50, 54-63, 67, 68, 70, 71, 73-79, 93, 102, 110, 119, 135, 136, 142-153

Sam68-like mammalian protein 1 (Slm1) 1, 2, 5, 11, 70, 73, 74, 93, 146-148
Sam68-like mammalian protein 2 (Slm2) 1, 2, 5, 11, 42, 70, 71, 146-148, 151
Schizophrenia 7, 16, 26, 33, 152
SH2 3, 54, 57, 58, 61, 73, 74, 148
SH3 3, 4, 16, 26, 27, 54, 55, 57, 58, 60, 61, 73, 74, 142, 143, 147, 148
SIAH1 73, 74
Signaling 3, 16, 25-33, 38, 42, 54, 55, 57-59, 61, 63, 70, 74, 78, 85-87, 89, 96, 97, 103, 109, 114, 118, 133, 136, 143, 147, 150
Signal transduction 1, 3-5, 11, 21, 25, 38, 54, 55, 57, 62, 63, 70, 83, 85, 93, 103, 142
Signal transduction and activator of RNA (STAR) 1-6, 8, 11, 20, 21, 25, 37-43, 47, 48, 50, 54-57, 59-64, 67, 70-79, 83, 85, 93-95, 99, 102-104, 106, 109, 110, 119, 120, 123, 135, 136, 142-145, 147-149, 151-153
Small nuclear ribonucleoprotein (snRNP) 124, 126-132
Small nuclear RNA (snRNA) 127, 129, 131, 132
Smy2 125, 133-136
Sperm 67-69, 76, 107
Spermatocyte 68, 69, 70, 74, 76-78, 147
Spermatogenesis 40, 67, 68-70, 76-79, 104, 106, 115, 145, 149
Spermatogonia 69, 78
Spliceosome cycle 123, 127, 128
Splicing factor 1 (SF1) 1-6, 11, 38, 43-50, 63, 94, 110, 117, 119, 123-127, 130-132, 135, 136, 145
Src family protein tyrosine kinase 4
SRp30c 78, 147
STAR protein 2, 4-6, 8, 21, 25, 37-39, 41-43, 48, 50, 54-57, 59-64, 67, 70, 71, 73, 74, 76-79, 93-95, 99, 102-104, 106, 110, 119, 142, 143, 145, 147-149, 151-153
Sub2 128, 130, 135, 136
Systematic evolution of ligands through exponential enrichment (SELEX) 39, 42, 49, 50, 78, 124, 145

T

Teratoma   109, 114
Testis   67-70, 74, 76-79, 104, 145, 149
Thermodynamics   37
Tissue differentiation   95, 98
Tra2β   78
Transition protein   76
Translation   6, 28, 29, 30, 41, 55, 56, 76,
      110, 113-120, 130, 133, 134, 136,
      147
Translational control   64, 67, 76, 77
Translational repression   110, 113-118,
      135
T-STAR   61, 67, 70-79
Tumorigenesis   5, 26, 27, 32, 33, 57,
      150, 153
Type V myosin   133, 134
Tyrosine kinase   3, 4, 8, 10, 14, 25, 30,
      31, 57-59, 63, 74, 85, 98, 143, 146,
      147, 149, 150, 152
Tyrosine phosphorylation   26, 33, 55,
      57-60, 62, 63, 74, 142, 143, 147,
      148, 150, 152

U

3'UTR   7, 9-19, 85, 134
U2AF   77, 126, 127
U2AF65   59, 63, 123-128, 130, 136
U2AF homology motif (UHM)   126, 127
UHM ligand motif (ULM)   125-127

V

Vascular endothelium   87
Vasculogenesis   84, 85, 89
Visceral endoderm   5, 82, 84-89

Y

Yolk sac   82, 84, 85, 87, 88
YT521B   78